HSC Year 12
MATHEMATICS EXTENSION 1

JOHN DRAKE

SERIES EDITOR: ROBERT YEN

· 2020 UPDATED SYLLABUS · 2020 UPDATED SYLLABUS · 2020 UPDATED SYLLABUS ·

A+

+ topic exams of HSC-style questions
+ practice HSC and mini-HSC exams
+ worked solutions with expert comments
+ HSC exam topic grids (2011–2020)

PRACTICE EXAMS

A+ HSC Mathematics Extension 1 Practice Exams

1st Edition

John Drake

ISBN 9780170459259

Publishers: Robert Yen, Kirstie Irwin

Project editor: Tanya Smith

Cover design: Nikita Bansal

Text design: Alba Design

Project designer: Nikita Bansal

Permissions researcher: Corrina Gilbert

Production controller: Karen Young

Typeset by: Nikki M Group Pty Ltd

Any URLs contained in this publication were checked for currency during the production process. Note, however, that the publisher cannot vouch for the ongoing currency of URLs.

NSW Education Standards Authority (NESA): 1995 Higher School Certificate Examination Mathematics 3 Unit (Additional) and 3/4 Unit (Common); Higher School Certificate Examination Mathematics Extension 1 (2001, 2002, 2019) © NSW Education Standards Authority for and on behalf of the Crown in right of the State of New South Wales.

For product information and technology assistance,
in Australia call **1300 790 853**;
in New Zealand call **0800 449 725**

For permission to use material from this text or product, please email **aust.permissions@cengage.com**

ISBN 978 0 17 045925 9

Cengage Learning Australia
Level 7, 80 Dorcas Street
South Melbourne, Victoria Australia 3205

Cengage Learning New Zealand
Unit 4B Rosedale Office Park
331 Rosedale Road, Albany, North Shore 0632, NZ

For learning solutions, visit **cengage.com.au**

Printed in China by 1010 Printing International Limited.
1 2 3 4 5 6 7 25 24 23 22 21

ABOUT THIS BOOK

Introducing *A+ HSC Year 12 Mathematics*, a new series of study guides designed to help students revise the topics of the new HSC maths courses and achieve success in their exams. *A+* is published by Cengage, the educational publisher of *Maths in Focus* and *New Century Maths*.

For each HSC maths course, Cengage has developed a STUDY NOTES book and a PRACTICE EXAMS book. These study guides have been written by experienced teachers who have taught the new courses, some of whom are involved in HSC exam marking and writing. This is the first study guide series to be published after the first HSC exams of the new courses in 2020, so it incorporates the latest changes to the syllabus and exam format

This book, *A+ HSC Year 12 Mathematics Extension 1 Practice Exams,* contains topic exams and practice HSC exams, both written and formatted in the style of the HSC exams, with spaces for students to write answers. Worked solutions are provided along with the author's expert comments and advice, including how each exam question is marked. An HSC exam topic grid (2011–2020) guides students to where and how each topic has been tested in past HSC exams.

Mathematics Extension 1 Year 12 topics

1. Mathematical induction

2. Vectors

3. Trigonometric equations

4. Further integration

5. Volumes and differential equations

6. The binomial distribution

This book contains:

- 6 topic exams: 1-hour mini-HSC exams on each topic + worked solutions

- 2 practice mini-HSC exams: 1-hour exams + worked solutions

- 2 practice HSC exams: full (3-hour) exams + worked solutions

- HSC exam reference sheet of formulas

- bonus: worked solutions to the 2020 HSC exam.

The companion A+ STUDY NOTES book contains topic summaries and graded practice questions, grouped into the same 6 broad topics, including for each topic a concept map, glossary and HSC exam topic grid.

Both books can be used for revision after a topic has been learned, as well as for preparation for the trial and HSC exams. Before you begin any questions, make sure you have a thorough understanding of the topic you will be undertaking.

9780170459259

CONTENTS

CHAPTER 1

MATHEMATICAL INDUCTION

CHAPTER 2

VECTORS

CHAPTER 3

TRIGONOMETRIC EQUATIONS

CHAPTER 4

FURTHER INTEGRATION

CHAPTER 5

VOLUMES AND DIFFERENTIAL EQUATIONS

CONTENTS

YEAR 12 COURSE OVERVIEW

MATHEMATICAL INDUCTION

Series proofs

Step 1:   Show true for $n = 1$

Step 2:   Assume true for $n = k$

Step 3:   Prove true for $n = k + 1$:

$$S_{k+1} = S_k + T_{k+1}$$

Step 4:   Conclusion: 'Hence the statement is true for all integers $n \geq 1$ by mathematical induction'.

Divisibility proofs

Step 1:   Show true for $n = 1$

Step 2:   Assume true for $n = k$ by making the statement over the divisor equal to M

Step 3:   Prove true for $n = k + 1$

Step 4:   Conclusion: 'Hence, the statement is true for all integers $n \geq 1$ by mathematical induction'.

VECTORS

Operations with vectors

- Magnitude
- Length and angle
- Adding and subtracting
- Scalar multiplication
- Scalar (dot) product
- Projections

Types of vectors

- Position vector
- Zero vector
- Parallel and perpendicular vectors
- Unit vector

Geometric descriptions

- Column and component form
- The angle between 2 vectors
- Displacement vector
- Geometrical proofs

Projectile motion

- Deriving equations
- Vector functions
- Time equations

TRIGONOMETRIC EQUATIONS

Using trigonometric identities

- Compound angle identities
- Double angle identities
- Products to sums identities
- t-formulas
- Solving trigonometric equations

Auxiliary angle method

- $a\cos x + b\sin x = c$

Proving trigonometric identities

FURTHER INTEGRATION

Integration by substitution

- Let $u = \ldots$
- Given the substitution
- Choosing the substitution
- Trigonometric substitutions

Trigonometric integrals

- Integrating $\sin^2 x$ and $\cos^2 x$
- Trigonometric identities
- Products to sums formulas

Inverse trigonometric functions

- Differentiating inverse functions
- Differentiating inverse trigonometric functions
- Integrals involving inverse functions

Areas of integration

- Areas about the x-axis
- Areas about the y-axis

Volumes of solids of revolution

- Volumes about the x-axis
- Volumes about the y-axis

DIFFERENTIAL EQUATIONS

Solving differential equations

$$\frac{dy}{dx} = f(x)$$

$$\frac{dy}{dx} = g(y)$$

$$\frac{dy}{dx} = f(x)\,g(y): \text{separation of variables}$$

Direction fields

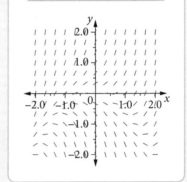

Application of differential equations

- Exponential growth and decay
- Newton's law of cooling
- Logistic equations

THE BINOMIAL DISTRIBUTION

Binomial distribution

- Bernoulli distributions
- Binomial distributions
- Mean and variance

$$E(X) = np$$

$$\text{Var}(X) = np(1-p)$$

Binomial probability

$$P(X = r) = {}^nC_r p^r (1-p)^{n-r}$$

$$X \sim \text{Bin}(n, p)$$

Sample proportions

- Sample proportion distributions

$$E(\hat{p}) = p$$

$$\text{Var}(\hat{p}) = \frac{pq}{n}$$

- Normal approximation

SYLLABUS REFERENCE GRID

Topic and subtopics	A+ HSC Year 12 Mathematics Extension 1 Practice Exams chapter
PROOF	
ME-P1 Proof by mathematical induction	1 Mathematical induction
VECTORS	
ME V1 Introduction to vectors V1.1 Introduction to vectors V1.2 Further operations with vectors V1.3 Projectile motion	2 Vectors
TRIGONOMETRIC FUNCTIONS	
ME-T3 Trigonometric equations	3 Trigonometric equations
CALCULUS	
ME-C2 Further calculus skills	4 Further integration
ME-C3 Applications of calculus C3.1 Further area and volumes of solids of revolution C3.2 Differential equations	5 Volumes and differential equations
STATISTICAL ANALYSIS	
ME-S1 The binomial distribution S1.1 Bernoulli and binomial distributions S1.2 Normal approximation for the sample proportion	6 The binomial distribution

ABOUT THE AUTHOR

John Drake teaches mathematics and computing at Parramatta Marist High School, Westmead, including online lessons for students across the Catholic Education Diocese of Parramatta. He is a regular presenter at teacher conferences, has been involved in HSC marking for computing, and is a contributor to MANSW's annual HSC mathematics exam solutions.

A+ DIGITAL FLASHCARDS

Revise key terms and concepts online with the A+ Flashcards. Each topic for this course has a deck of digital flashcards you can use to test your understanding and recall. Just scan the QR code or type the URL into your browser to access them.

Note: You will need to create a free *NelsonNet* account.

https://get.ga/a-hsc-maths-ext-1

HSC EXAM FORMAT

Mathematics Extension 1 students complete two HSC exams: **Mathematics Advanced** and **Mathematics Extension 1**.

The following information about the exams was correct at the time of printing in 2021. Please check the NESA website in case it has changed. Visit www.educationstandards.nsw.edu.au, select 'Year 11–Year 12', 'Syllabuses A–Z', 'Mathematics Advanced/Extension 1', then 'Assessment and Reporting'. Scroll down to 'HSC examination specifications'.

Mathematics Advanced HSC exam

	Questions	Marks	Recommended time
Section I	10 multiple-choice questions Mark answers on the multiple-choice answer sheet.	10	15 min
Section II	Approx. 24 short-answer questions, including 2 or more questions worth 4 or 5 marks. Write answers on the lines provided on the paper.	90	2 h 45 min
Total		100	3 h

- Reading time: 10 minutes; use this time to preview the whole exam.

- Working time: 3 hours

- Questions focus on Year 12 outcomes but Year 11 knowledge may be examined.

- Answers are to be written on the question paper.

- A reference sheet is provided at the back of the exam paper and also in this book, containing common formulas.

- Common questions with the Mathematics Standard 2 HSC exam: 20–25 marks

- The 4- or 5-mark questions are usually complex problems that require many steps of working and careful planning.

- To help you plan your time, the halfway point of Section II is marked by a notice at the bottom of the relevant page; for example, 'Questions 11–23 are worth 46 marks in total'.

- Having 3 hours for a total of 100 marks means that you have an average of 1.8 minutes per mark (or approximately 5 minutes for 3 marks).

- If you budget 15 minutes for Section I, and 1 hour 15 minutes for each half of Section II, you will have 15 minutes at the end to check over your work and/or complete questions you missed.

Mathematics Extension 1 HSC exam

	Questions	Marks	Recommended time
Section I	10 multiple-choice questions	10	15 min
Section II	4 multi-part short-answer questions, average 15 marks each, including questions worth 4 or 5 marks	60	1 h 45 min
Total		70	2 h

- Reading time: 10 minutes; Working time: 2 hours.

- Answers are to be written in separate answer booklets.

- Having 2 hours for a total of 70 marks means that you have an average of 1.7 minutes per mark (or approximately 5 minutes for 3 marks).

- If you budget 15 minutes for Section I, then 20 minutes per question for Section II, you will have 25 minutes at the end to check over your work and/or complete questions you missed.

STUDY AND EXAM ADVICE

A journey of a thousand miles begins with a single step.

Lao Tzu (c. 570–490 BCE), Chinese philosopher

I've always believed that if you put in the work, the results will come.

Michael Jordan (1963–), American basketball player

Four PRACtical steps for maths study

1. **P**ractise your maths

- Do your homework.

- Learning maths is about mastering a collection of skills.

- You become successful at maths by doing it more, through regular practice and learning.

- Aim to achieve a high level of understanding.

2. **R**ewrite your maths

- Homework and study are not the same thing. Study is your private 'revision' work for strengthening your understanding of a subject.

- Before you begin any questions, make sure you have a thorough understanding of the topic.

- Take ownership of your maths. Rewrite the theory and examples in your own words.

- Summarise each topic to see the 'whole picture' and know it all.

3. **A**ttack your maths

- All maths knowledge is interconnected. If you don't understand one topic fully, then you may have trouble learning another topic.

- Mathematics is not an HSC course you can learn 'by halves' – you have to know it all!

- Fill in any gaps in your mathematical knowledge to see the 'whole picture'.

- Identify your areas of weakness and work on them.

- Spend most of your study time on the topics you find difficult.

4. **C**heck your maths

- After you have mastered a maths skill, such as graphing a quadratic equation, no further learning or reading is needed, just more practice.

- Compared to other subjects, the types of questions asked in maths exams are conventional and predictable.

- Test your understanding with revision exercises, practice papers and past exam papers.

- Develop your exam technique and problem-solving skills.

- Go back to steps 1–3 to improve your study habits.

Topic summaries and concept maps

Summarise each topic when you have completed it, to create useful study notes for revising the course, especially before exams. Use a notebook or folder to list the important ideas, formulas, terminology and skills for each topic. This book is a good study guide, but educational research shows that effective learning takes place when you rewrite learned knowledge in your own words.

A good topic summary runs for 2 to 4 pages. It is a condensed, personalised version of your course notes. This is your interpretation of a topic, so include your own comments, symbols, diagrams, observations and reminders. Highlight important facts using boxes and include a glossary of key words and phrases.

A concept map or mind map is a topic summary in graphic form, with boxes, branches and arrows showing the connections between the main ideas of the topic. This book contains examples of concept maps. The topic name is central to the map, with key concepts or subheadings listing important details and formulas. Concept maps are powerful because they present an overview of a topic on one large sheet of paper. Visual learners absorb and recall information better when they use concept maps.

When compiling a topic summary, use your class notes, your textbook and the A+ Study Notes book that accompanies this book. Ask your teacher for a copy of the course syllabus or the school's teaching program, which includes the objectives and outcomes of every topic in dot point form.

Attacking your weak areas

Most of your study time should be spent on attacking your weak areas to fill in any gaps in your maths knowledge. Don't spend too much time on work you already know well, unless you need a confidence boost! Ask your teacher, or use this book or your textbook to improve the understanding of your weak areas and to practise maths skills. Use your topic summaries for general revision, but spend longer study periods on overcoming any difficulties in your mastery of the course.

Practising with past exam papers

Why is practising with past exam papers such an effective study technique? It allows you to become familiar with the format, style and level of difficulty expected in an HSC exam, as well as the common topic areas tested. Knowing what to expect helps alleviate exam anxiety. Remember, mathematics is a subject in which the exam questions are fairly predictable. The exam writers are not going to ask many unusual questions. By the time you have worked through many past exam papers, this year's HSC exams won't seem that much different.

Don't throw your old exam papers away. Use them to identify your mistakes and weak areas for further study. Revising topics and then working on mixed questions is a great way to study maths. You might like to complete a past HSC exam paper under timed conditions to improve your exam technique.

Past HSC exam papers are available at the NESA website: visit www.educationstandards.nsw.edu.au and select 'Year 11 – Year 12', 'HSC exam papers'. NESA marking feedback and guidelines can also be viewed there. You can find past HSC exam papers with solutions online, in bookstores, at the Mathematical Association of NSW (www.mansw.nsw.edu.au) and at your school (ask your teacher) or library.

Preparing for an exam

- Make a study plan early; don't leave it until the last minute.
- Read and revise your topic summaries.
- Work on your weak areas and learn from your mistakes.
- Don't spend too much time studying concepts you know already.
- Revise by completing revision exercises and past exam papers or assignments.
- Vary the way you study so that you don't become bored: ask someone to quiz you, voice-record your summary, design a poster or concept map, or explain a concept to someone.
- Anticipate the exam:
  - How many questions will there be?
  - What are the types of questions: multiple-choice, short-answer, long-answer, problem-solving?
  - Which topics will be tested?
  - How many marks are there in each section?
  - How long is the exam?
  - How much time should I spend on each question/section?
  - Which formulas are on the reference sheet and how do I use them in the exam?

During an exam

1. Bring all of your equipment, including a ruler and calculator (check that your calculator works and is in RADIANS mode for trigonometric functions and DEGREES for trigonometric measurements). A highlighter pen may help for tables, graphs and diagrams.

2. Don't worry if you feel nervous before an exam – this is normal and can help you to perform better; however, being too casual or too anxious can harm your performance. Just before the exam begins, take deep, slow breaths to reduce any stress.

3. Write clearly and neatly in black or blue pen, not red. Use a pencil only for diagrams and constructions.

4. Use the **reading time** to browse through the exam to see the work that is ahead of you and the marks allocated to each question. Doing this will ensure you won't miss any questions or pages. Note the harder questions and allow more time for working on them. Leave them if you get stuck, and come back to them later.

5. Attempt every question. It is better to do most of every question and score some marks, rather than ignore questions completely and score 0 for them. Don't leave multiple-choice questions unanswered! Even if you guess, you have a chance of being correct.

6. Easier questions are usually at the beginning, with harder ones at the end. Do an easy question first to boost your confidence. Some students like to leave multiple-choice questions until last so that, if they run out of time, they can make quick guesses. However, some multiple-choice questions can be quite difficult.

7. Read each question and identify what needs to be found and what topic/skill it is testing. The number of marks indicates how much time and working out is required. Highlight any important keywords or clues. Do you need to use the answer to the previous part of the question?

8. After reading each question, and before you start writing, spend a few moments planning and thinking.

9. You don't need to be writing all of the time. What you are writing may be wrong and a waste of time. Spend some time considering the best approach.

10. Make sure each answer seems reasonable and realistic, especially if it involves money or measurement.

11. Show all necessary working, write clearly, draw big diagrams, and set your working out neatly. Write solutions to each part underneath the previous step so that your working out goes down the page, not across.

12. Use a ruler to draw (or read) half-page graphs with labels and axes marked, or to measure scale diagrams.

13. Don't spend too much time on one question. Keep an eye on the time.

14. Make sure you have answered the question. Did you remember to round the answer and/or include units? Did you use all of the relevant information given?

15. If a hard question is taking too long, don't get bogged down. If you're getting nowhere, retrace your steps, start again, or skip the question (circle it) and return to it later with a clearer mind.

16. If you make a mistake, cross it out with a neat line. Don't scribble over it completely or use correction fluid or tape (which is time-consuming and messy). You may still score marks for crossed-out work if it is correct, but don't leave multiple answers! Keep track of your answer booklets and ask for more writing paper if needed.

17. Don't cross out or change an answer too quickly. Research shows that often your first answer is the correct one.

18. Don't round your answer in the middle of a calculation. Round at the end only.

19. Be prepared to write words and sentences in your answers, but don't use abbreviations that you've just made up. Use correct terminology and write one or two sentences for 2 or 3 marks, not mini-essays.

20. If you have time at the end of the exam, double-check your answers, especially for the more difficult questions or questions you are uncertain about.

Ten exam habits of the best HSC students

1. Has clear and careful working and checks their answers

2. Has a strong understanding of basic algebra and calculation

3. Reads (and answers) the whole question

4. Chooses the simplest and quickest method

5. Checks that their answer makes sense or sounds reasonable

6. Draws big, clear diagrams with details and labels

7. Uses a ruler for drawing, measuring and reading graphs

8. Can explain answers in words when needed, in one or two clear sentences

9. Uses the previous part/s of a question to solve the next part of the question

10. Rounds answers at the end, not before.

Further resources

Visit the NESA website (www.educationstandards.nsw.edu.au) for the following resources. Select 'Year 11 – Year 12' and then 'Syllabuses A–Z' or 'HSC exam papers'.

- Mathematics Advanced and Extension 1 syllabuses

- Past HSC exam papers, including marking feedback and guidelines

- Sample HSC questions/exam papers and marking guidelines

Before 2020, 'Mathematics Advanced' was called 'Mathematics' and although 'Mathematics Extension 1' had the same name, it was a different course with some topics that no longer exist. For these exam papers, select 'Year 11 – Year 12', 'Resources archive', 'HSC exam papers archive'.

MATHEMATICAL VERBS

A glossary of 'doing words' common in maths problems and HSC exams

analyse
study in detail the parts of a situation

apply
use knowledge or a procedure in a given situation

calculate
See **evaluate**

classify/identify
state the type, name or feature of an item or situation

comment
express an observation or opinion about a result

compare
show how two or more things are similar or different

complete
fill in detail to make a statement, diagram or table correct or finished

construct
draw an accurate diagram

convert
change from one form to another, for example, from a fraction to a decimal, or from kilograms to grams

decrease
make smaller

describe
state the features of a situation

estimate
make an educated guess for a number, measurement or solution, to find roughly or approximately

evaluate/calculate
find the value of a numerical expression, for example, 3×8^2 or $4x + 1$ when $x = 5$

expand
remove brackets in an algebraic expression, for example, expanding $3(2y + 1)$ gives $6y + 3$

explain
describe why or how

give reasons
show the rules or thinking used when solving a problem. *See also* **justify**

graph
display on a number line, number plane or statistical graph

hence find/prove
calculate an answer or demonstrate a result using previous answers or information supplied

identify
See **classify**

increase
make larger

interpret
find meaning in a mathematical result

justify
give reasons or evidence to support your argument or conclusion. *See also* **give reasons**

measure
determine the size of something, for example, using a ruler to determine the length of a pen

prove
See **show/prove that**

recall
remember and state

show/prove that
(in questions where the answer is given) use calculation, procedure or reasoning to demonstrate that an answer or result is true

simplify
express a result such as a ratio or algebraic expression in its most basic, shortest, neatest form

sketch
draw a rough diagram that shows the general shape or ideas (less accurate than **construct**)

solve
calculate the value(s) of an unknown pronumeral in an equation or inequality

state
See **write**

substitute
replace a variable with a number

verify
check that a solution or result is correct, usually by substituting back into an equation or referring back to the problem

write/state
give an answer, formula or result without showing any working or explanation (This usually means that the answer can be found mentally, or in one step)

SYMBOLS AND ABBREVIATIONS

$=$	is equal to
$\neq$	is not equal to
$\approx$	is approximately equal to
$<$	is less than
$>$	is greater than
$\leq$	is less than or equal to
$\geq$	is greater than or equal to
( )	parentheses, round brackets
[ ]	(square) brackets
{ }	braces
$\pm$	plus or minus
π	pi = 3.14159 ...
$\equiv$	is congruent/identical to
$\circ$	degree
$\angle$	angle
Δ	triangle, the discriminant
$\parallel$	is parallel to
$\perp$	is perpendicular to
x^2	x squared, $x \times x$
x^3	x cubed, $x \times x \times x$
$\cup$	union
$\cap$	intersection
∞	infinity
$\lvert x \rvert$	absolute value or magnitude of x
$\underset{\sim}{v}$	the vector v
$\overrightarrow{AB}$	the vector AB
$\underset{\sim}{u} \cdot \underset{\sim}{v}$	the scalar product of $\underset{\sim}{u}$ and $\underset{\sim}{v}$
$\text{proj}_{\underset{\sim}{u}} \underset{\sim}{v}$	the projection of $\underset{\sim}{v}$ onto $\underset{\sim}{u}$
$\lim\limits_{h \to 0}$	the limit as $h \to 0$
$\dfrac{dy}{dx}, y', f'(x)$	the first derivative of $y, f(x)$
$\dfrac{d^2 y}{dx^2}, y'', f''(x)$	the second derivative of $y, f(x)$
$\int f(x)\,dx$	the integral of $f(x)$
$f^{-1}(x)$	the inverse function of $f(x)$
$\sin^{-1}, \arcsin$	the inverse sine function
Σ	sigma, the sum of
$\therefore$	therefore

$[a, b], a \leq x \leq b$	the interval of x-values from a to b (including a and b)
$(a, b), a < x < b$	the interval of x-values between a and b (excluding a and b)
$P(E)$	the probability of event E occurring
$P(\bar{E})$	the probability of event E not occurring
$A \cup B$	A union B, A or B
$A \cap B$	A intersection B, A and B
$P(A \mid B)$	the probability of A given B
$n!$	n factorial, $n(n-1)(n-2) \ldots \times 1$
$^nC_r, \dbinom{n}{r}$	the number of combinations of r objects from n objects
^nP_r	the number of permutations of r objects from n objects
PDF	probability density function
CDF	cumulative distribution function
$X \sim \text{Bin}(n, p)$	X is a random variable of the binomial distribution
$\hat{p}$	sample proportion
LHS	left-hand side
RHS	right-hand side
p.a.	per annum (per year)
cos	cosine ratio
sin	sine ratio
tan	tangent ratio
$\bar{x}$	the mean
$\mu = E(X)$	the population mean, expected value
σ	the standard deviation
$\text{Var}(X) = \sigma^2$	the variance
Q_1	first quartile or lower quartile
Q_2	median (second quartile)
Q_3	third quartile or upper quartile
IQR	interquartile range
α	alpha
θ	theta
m	gradient
RTP	required to prove

9780170459259

A+ HSC YEAR 12 MATHEMATICS

STUDY NOTES

Authors:
Tania Eastcott
Rachel Eastcott

Sarah Hamper

Karen Man
Ashleigh Della Marta

Jim Green
Janet Hunter

PRACTICE EXAMS

Authors:
Adrian Kruse

Simon Meli

John Drake

Jim Green
Janet Hunter

9780170459259

CHAPTER 1
TOPIC EXAM

Mathematical induction

ME-P1 Proof by mathematical induction

• A reference sheet is provided on page 155 at the back of this book. • For questions in Section II, show relevant mathematical reasoning and/or calculations.	**Reading time: 5 minutes** **Working time: 1 hour** **Total marks: 35**

Section I – 5 questions, 5 marks
- Attempt Questions 1–5
- Allow about 8 minutes for this section

Section II – 2 questions, 30 marks
- Attempt Questions 6–7
- Allow about 52 minutes for this section

Section I

• Attempt Questions 1–5 • Allow about 8 minutes for this section	**5 marks**

Question 1

Let $P(n)$ be the statement $2 + 4 + 6 + \cdots + 2n = n(n + 1)$.

What is $P(1)$?

A $n = 1$

B $2 = 1(1 + 1)$

C $2 + 4 + 6 + \cdots + 2k = k(k + 1)$

D $2 + 4 + 6 + \cdots + 2n = n(n + 1)$

Question 2

Eddie wants to prove by mathematical induction that $2 + 4 + 6 + \cdots + 2n = n(n + 1)$ for all integers $n \geq 1$. Let $P(n)$ be the statement $2 + 4 + 6 + \cdots + 2n = n(n + 1)$.

Which is the inductive step?

A $P(1)$ is true

B $P(k)$ is true

C If $P(1)$ is true, then $P(k)$ is true

D If $P(k)$ is true, then $P(k + 1)$ is true

Question 3

Eddie wants to prove by mathematical induction that $2 + 4 + 6 + \cdots + 2n = n(n + 1)$ for all integers $n \geq 1$.

What does Eddie need to prove?

A $k(k + 1) + 2k + 1 = k(k + 2)$

B $k(k + 1) + 2k = k(k + 1) + 1$

C $k(k + 1) + (k + 1) = (k + 1)^2$

D $k(k + 1) + 2(k + 1) = (k + 1)(k + 2)$

Question 4

Freda wants to write that an integer n is divisible by 3.

What should she write?

A $n = 3$

B $n = 3M$ for some integer M

C $n = \dfrac{M}{3}$ for some integer M

D $n = M + 3$ for some integer M

Question 5

When proving by mathematical induction, which step is optional?

A The initial case

B The inductive step

C The conclusion

D None of the above

Section II

• Attempt Questions 6–7 **30 marks**
• Allow about 52 minutes for this section
• Answer the questions in the spaces provided. These spaces provide guidance for the expected length of response.
• Your responses should include relevant mathematical reasoning and/or calculations.

Question 6 (15 marks)

a   Prove by mathematical induction that for all integers $n \geq 1$,   3 marks

$$2 \times 3 + 4 \times 6 + 6 \times 9 + \cdots + 2n \times 3n = n(n + 1)(2n + 1).$$

b Prove by mathematical induction that $7^n - 1$ is divisible by 6, for any integer $n \geq 1$.   3 marks

c Eddie wants to prove by mathematical induction that, for all integers $n \geq 1$,

$$1 + 3 + 5 + \cdots + (2n - 1) = (n - 1)(n + 1).$$

i Show that if the above statement is true for $n = k$, then it is also true for $n = k + 1$.   2 marks

ii Explain why proof by mathematical induction fails here. *1 mark*

d Prove by mathematical induction that, for all integers $n \geq 1$, *3 marks*

$$\frac{1}{4} + \frac{1}{28} + \frac{1}{70} + \cdots + \frac{1}{(3n-2)(3n+1)} = \frac{n}{3n+1}.$$

e Prove by mathematical induction that $n^3 + 5n$ is divisible by 3, for any integer $n \geq 1$.   3 marks

Question 7 (15 marks)

a Prove by mathematical induction that $3^{3n-1} + 5^{3n-2}$ is divisible by 7, for any integer $n \geq 1$.   3 marks

b   **i**   Show that $2 \cos(\alpha + \beta) \sin \beta = \sin(\alpha + 2\beta) - \sin \alpha$.   1 mark

    ii   Hence, prove by mathematical induction that, for all integers $n \geq 1$,   3 marks

$$\cos x + \cos 3x + \cos 5x + \cdots + \cos(2n-1)x = \frac{\sin 2nx}{2 \sin x}.$$

c i Show that $x^3 + 4x^2 + 3x - 2 = (x^2 + 2x - 1)(x + 2)$.   1 mark

ii Hence, prove by mathematical induction that, for all integers $n \geq 1$,   3 marks

$$0 \times 1! + 7 \times 2! + 26 \times 3! + \cdots + (n^3 - 1) \times n! = (n^2 - 2)(n + 1)! + 2.$$

d i Show that $\binom{n}{r} + \binom{n}{r+1} = \binom{n+1}{r+1}$ for positive integers r and n with $r < n$.

1 mark

ii Hence, prove by mathematical induction that, for all positive integers r and n,

3 marks

$$\binom{r}{r} + \binom{r+1}{r} + \binom{r+2}{r} + \cdots + \binom{r+n-1}{r} = \binom{r+n}{r+1}.$$

END OF PAPER

WORKED SOLUTIONS

Section I (1 mark each)

Question 1

B Substitute $n = 1$ into $P(n)$ to get $2 = 1(1 + 1)$.

This is an easy question if you understand Step 1 in an induction series proof.

Question 2

D

This is an 'understanding' question.

Question 3

D Substitute $n = k + 1$ into the statement and substitute the assumption $P(k)$ is true.

This is a straightforward question if you understand Step 3 in an induction series proof.

Question 4

B

Question 5

D

Questions 4 and 5 are both 'understanding' questions.

Section II (✓ = 1 mark)

Question 6 (15 marks)

a Let $P(n)$ be the statement $2 \times 3 + 4 \times 6 + 6 \times 9 + \cdots + 2n \times 3n = n(n + 1)(2n + 1)$.

$P(1)$ is $2 \times 3 = 1(1 + 1)[2(1) + 1]$.

LHS $= 6$, RHS $= 6 =$ LHS.

$\therefore P(1)$ is true. ✓

Assume $P(k)$ is true.

$P(k)$ is $2 \times 3 + 4 \times 6 + 6 \times 9 + \cdots + 2k \times 3k = k(k + 1)(2k + 1)$.   [*]

$P(k + 1)$ is $2 \times 3 + 4 \times 6 + 6 \times 9 + \cdots + 2k \times 3k + 2(k + 1) \times 3(k + 1) = (k + 1)(k + 1 + 1)[2(k + 1) + 1]$.

RHS $= (k + 1)(k + 2)(2k + 3)$

$$\begin{aligned}
\text{LHS} &= k(k + 1)(2k + 1) + 2(k + 1) \times 3(k + 1) \quad \text{by [*]} \checkmark \\
&= (k + 1)[k(2k + 1) + 6(k + 1)] \\
&= (k + 1)(2k^2 + k + 6k + 6) \\
&= (k + 1)(2k^2 + 7k + 6) \\
&= (k + 1)(k + 2)(2k + 3) \\
&= \text{RHS}
\end{aligned}$$

$\therefore P(k + 1)$ is true.

So by mathematical induction, $P(n)$ is true for all integers $n \geq 1$. ✓

This is a straightforward induction series proof. In the inductive step, avoid expanding the polynomial because factorising polynomials is difficult and time consuming.

b Let $P(n)$ be the statement $7^n - 1$ is divisible by 6.

$P(1)$ is $7^1 - 1$ is divisible by 6.

$7^1 - 1 = 6$, which is divisible by 6.

$\therefore P(1)$ is true. ✓

Assume $P(k)$ is true.   [*]

$P(k)$ is $7^k - 1$ is divisible by 6.

Thus, $7^k - 1 = 6p$ for some integer p.

$\therefore 7^k = 6p + 1$ [*]

$P(k + 1)$ is $7^{k+1} - 1$ is divisible by 6.

$$\begin{aligned}
7^{k+1} - 1 &= 7(7^k) - 1 \\
&= 7(6p + 1) - 1 \quad \text{by [*]} \checkmark \\
&= 42p + 7 - 1 \\
&= 42p + 6 \\
&= 6(7p + 1), \text{ which is divisible by 6.}
\end{aligned}$$

$\therefore P(k + 1)$ is true.

So by mathematical induction, $P(n)$ is true for all integers $n \geq 1$. ✓

This is a straightforward induction divisibility proof.

c i Assume $1 + 3 + 5 + \cdots + (2k - 1)$
$= (k - 1)(k + 1)$.   [*]

When $n = k + 1$,

$$\begin{aligned}
\text{LHS} \\
&= 1 + 3 + 5 + \cdots + (2k - 1) + [2(k + 1) - 1] \\
&= (k - 1)(k + 1) + [2k + 2 - 1] \quad \text{by [*]} \checkmark \\
&= k^2 - 1 + 2k + 1 \\
&= k^2 + 2k \\
&= k(k + 2)
\end{aligned}$$

$$\begin{aligned}
\text{RHS} &= (k + 1 - 1)(k + 1 + 1) \\
&= k(k + 2) \\
&= \text{LHS}
\end{aligned}$$

So it is true for $n = k + 1$. ✓

Be careful to only use the assumption.

ii When $n = 1$,

LHS $= 1$

RHS $= (1 - 1)(1 + 1) = 0 \neq$ LHS

So the proof fails because the initial case $P(1)$ is not true. ✓

d Let $P(n)$ be the statement $\dfrac{1}{4} + \dfrac{1}{28} + \dfrac{1}{70} + \cdots + \dfrac{1}{(3n-2)(3n+1)} = \dfrac{n}{3n+1}$.

$P(1)$ is $\dfrac{1}{4} = \dfrac{1}{3(1)+1}$.

LHS $= \dfrac{1}{4}$, RHS $= \dfrac{1}{4} =$ LHS.

$\therefore P(1)$ is true. ✓

Assume $P(k)$ is true.

$P(k)$ is $\dfrac{1}{4} + \dfrac{1}{28} + \dfrac{1}{70} + \cdots + \dfrac{1}{(3k-2)(3k+1)} = \dfrac{k}{3k+1}$. [*]

$P(k+1)$ is $\dfrac{1}{4} + \dfrac{1}{28} + \dfrac{1}{70} + \cdots + \dfrac{1}{(3k-2)(3k+1)} + \dfrac{1}{(3(k+1)-2)(3(k+1)+1)} = \dfrac{k+1}{3(k+1)+1}$

RHS $= \dfrac{k+1}{3k+4}$

$\begin{aligned}
\text{LHS} &= \dfrac{k}{3k+1} + \dfrac{1}{(3(k+1)-2)(3(k+1)+1)} \quad \text{by [*]} ✓ \\
&= \dfrac{k}{3k+1} + \dfrac{1}{(3k+1)(3k+4)} \\
&= \dfrac{k(3k+4)}{(3k+1)(3k+4)} + \dfrac{1}{(3k+1)(3k+4)} \\
&= \dfrac{3k^2+4k+1}{(3k+1)(3k+4)} \\
&= \dfrac{(3k+1)(k+1)}{(3k+1)(3k+4)} \\
&= \dfrac{k+1}{3k+4} \\
&= \text{RHS}
\end{aligned}$

$\therefore P(k+1)$ is true.

So by mathematical induction, $P(n)$ is true for all integers $n \geq 1$. ✓

> There is a lot of detailed algebra here. Be very careful with expanding, simplifying and adding fractions.

e Let $P(n)$ be the statement $n^3 + 5n$ is divisible by 3.

$P(1)$ is $1^3 + 5(1)$ is divisible by 3.

$1^3 + 5(1) = 6$, which is divisible by 3.

$\therefore P(1)$ is true. ✓

Assume $P(k)$ is true. [*]

$P(k)$ is $k^3 + 5k$ is divisible by 3.

Thus, $k^3 + 5k = 3p$ for some integer p.

$\therefore k^3 = 3p - 5k$ [*]

$P(k+1)$ is $(k+1)^3 + 5(k+1)$ is divisible by 3.

$\begin{aligned}
(k+1)^3 &+ 5(k+1) \\
&= k^3 + 3k^2 + 3k + 1 + 5k + 5 \\
&= k^3 + 3k^2 + 8k + 6 \\
&= 3p - 5k + 3k^2 + 8k + 6 \quad \text{by [*]} ✓ \\
&= 3p + 3k^2 + 3k + 6 \\
&= 3(p + k^2 + k + 2), \text{ which is divisible by 3.}
\end{aligned}$

$\therefore P(k+1)$ is true.

So by mathematical induction, $P(n)$ is true for all integers $n \geq 1$. ✓

> Be careful with expanding and substituting the inductive step.

Question 7 (15 marks)

a Let $P(n)$ be the statement $3^{3n-1} + 5^{3n-2}$ is divisible by 7.

$P(1) = 3^{3(1)-1} + 5^{3(1)-2}$ is divisible by 7.

$$3^{3(1)-1} + 5^{3(1)-2} = 3^2 + 5^1$$
$$= 14, \text{ which is divisible by 7.}$$

$\therefore P(1)$ is true. ✓

Assume $P(k)$ is true.   [*]

$P(k)$ is $3^{3k-1} + 5^{3k-2}$ is divisible by 7.

Thus, $3^{3k-1} + 5^{3k-2} = 7p$ for some integer p.
$$\therefore 3^{3k-1} = 7p - 5^{3k-2} \quad [*]$$

$P(k+1)$ is $3^{3(k+1)-1} + 5^{3(k+1)-2}$ is divisible by 7.

$$3^{3(k+1)-1} + 5^{3(k+1)-2} = 3^{3k+2} + 5^{3k+1}$$
$$= 3^{3k-1+3} + 5^{3k+1}$$
$$= (7p - 5^{3k-2})\,3^3 + 5^{3k+1} \quad \text{by [*]} \;✓$$
$$= 27(7p - 5^{3k-2}) + 5^{3k-2+3}$$
$$= 189p - 27(5^{3k-2}) + 5^{3k-2}(5^3)$$
$$= 189p - 27(5^{3k-2}) + 125(5^{3k-2})$$
$$= 189p + 98(5^{3k-2})$$
$$= 7[27p + 14(5^{3k-2})], \text{ which is divisible by 7.}$$

$\therefore P(k+1)$ is true.

So by mathematical induction, $P(n)$ is true for all integers $n \geq 1$. ✓

There are other ways to do this more complex divisibility induction proof, but they depend on completing the inductive step and manipulating exponents (powers) correctly.

b i $2\cos(\alpha + \beta)\sin\beta = 2 \times \dfrac{1}{2}[\sin(\alpha + \beta + \beta) - \sin(\alpha + \beta - \beta)]$ (from the reference sheet)

$$= \sin(\alpha + 2\beta) - \sin\alpha \;✓$$

This is easy if you know what formulas are on the HSC exam reference sheet. Refer to the back of this book.

ii Let $P(n)$ be the statement $\cos x + \cos 3x + \cos 5x + \cdots + \cos(2n-1)x = \dfrac{\sin 2nx}{2\sin x}$.

$P(1)$ is $\cos x = \dfrac{\sin 2x}{2\sin x}$.

$\text{RHS} = \dfrac{2\sin x \cos x}{2\sin x} = \cos x = \text{LHS}$

$\therefore P(1)$ is true. ✓

Assume $P(k)$ is true.

$P(k)$ is $\cos x + \cos 3x + \cos 5x + \cdots + \cos(2k-1)x = \dfrac{\sin 2kx}{2\sin x}$.   [*]

$P(k+1)$ is $\cos x + \cos 3x + \cos 5x + \cdots + \cos(2k-1)x + \cos[(2(k+1)-1)x] = \dfrac{\sin 2(k+1)x}{2\sin x}$.

$$\text{LHS} = \frac{\sin 2kx}{2\sin x} + \cos(2k+1)x \quad \text{by [*]} \checkmark$$

$$= \frac{\sin 2kx}{2\sin x} + \frac{2\cos(2k+1)x\sin x}{2\sin x}$$

$$= \frac{\sin 2kx + \sin(2k+2)x - \sin 2kx}{2\sin x} \quad \text{(from part \textbf{i})}$$

$$= \frac{\sin 2(k+1)x}{2\sin x}$$

$$= \text{RHS}$$

$\therefore P(k+1)$ is true.

So by mathematical induction, $P(n)$ is true for all integers $n \geq 1$. $\checkmark$

This is a more challenging induction sum proof, using the formula from part **i**.

c i $\text{RHS} = x^3 + 2x^2 - x + 2x^2 + 4x - 2$

$\qquad = x^3 + 4x^2 + 3x - 2$

$\qquad = \text{LHS} \checkmark$

Don't forget that you can prove that RHS = LHS (expanding), which is easier than LHS = RHS (factorising, polynomial division).

ii Let $P(n)$ be the statement $0 \times 1! + 7 \times 2! + 26 \times 3! + \cdots + (n^3 - 1) \times n! = (n^2 - 2)(n+1)! + 2$.

$P(1)$ is $0 \times 1! = (1^2 - 2)(1+1)! + 2$.

$\text{LHS} = 0$

$\text{RHS} = -1 \times 2! + 2$

$\qquad = 0$

$\qquad = \text{LHS}$

$\therefore P(1)$ is true. $\checkmark$

Assume $P(k)$ is true.

$P(k)$ is $0 \times 1! + 7 \times 2! + 26 \times 3! + \cdots + (k^3 - 1) \times k! = (k^2 - 2)(k+1)! + 2.$ [*]

$P(k+1)$ is $0 \times 1! + \cdots + (k^3 - 1) \times k! + [(k+1)^3 - 1](k+1)! = [(k+1)^2 - 2](k+1+1)! + 2.$

$\text{RHS} = (k^2 + 2k - 1)(k+2)! + 2$

$\text{LHS} = (k^2 - 2)(k+1)! + 2 + (k^3 + 3k^2 + 3k + 1 - 1)(k+1)! \quad \text{by [*]} \checkmark$

$\qquad = (k^2 - 2)(k+1)! + 2 + (k^3 + 3k^2 + 3k)(k+1)!$

$\qquad = (k+1)!(k^2 - 2 + k^3 + 3k^2 + 3k) + 2$

$\qquad = (k+1)!(k^3 + 4k^2 + 3k - 2) + 2$

$\qquad = (k+1)!(k^2 + 2k - 1)(k+2) + 2 \quad \text{(from part \textbf{i})}$

$\qquad = (k^2 + 2k - 1)(k+2)(k+1)! + 2$

$\qquad = (k^2 + 2k - 1)(k+2)! + 2$

$\qquad = \text{RHS}$

$\therefore P(k+1)$ is true.

So by mathematical induction, $P(n)$ is true for all integers $n \geq 1$. $\checkmark$

This is the same format as all induction series proofs, but you need to be very careful with the algebra and know how to manipulate factorials. Always look to use the solution for part **i**.

d i LHS $= \dfrac{n!}{r!(n-r)!} + \dfrac{n!}{(r+1)!(n-r-1)!}$ (from the reference sheet)

$= \dfrac{n!(r+1)}{(r+1)!(n-r)!} + \dfrac{n!(n-r)}{(r+1)!(n-r)!}$

$= \dfrac{n!(r+1+n-r)}{(r+1)!(n-r)!}$

$= \dfrac{n!(n+1)}{(r+1)!(n-r)!}$

$= \dfrac{(n+1)!}{(r+1)!(n+1-r-1)!}$

$= \dbinom{n+1}{r+1}$ (from the reference sheet)

$= \text{RHS}$ ✓

> This proof of the Pascal's triangle identity is a common question. Proficiency with manipulating expressions with factorials and knowing the combinations/coefficients formula is crucial here.

ii Let $P(n)$ be the statement $\dbinom{r}{r} + \dbinom{r+1}{r} + \dbinom{r+2}{r} + \cdots + \dbinom{r+n-1}{r} = \dbinom{r+n}{r+1}$.

$P(1)$ is $\dbinom{r}{r} = \dbinom{r+1}{r+1}$.

$\text{LHS} = 1$, $\text{RHS} = 1 = \text{LHS}$

$\therefore P(1)$ is true. ✓

Assume $P(k)$ is true.

$P(k)$ is $\dbinom{r}{r} + \dbinom{r+1}{r} + \dbinom{r+2}{r} + \cdots + \dbinom{r+k-1}{r} = \dbinom{r+k}{r+1}$.   [*]

$P(k+1)$ is $\dbinom{r}{r} + \dbinom{r+1}{r} + \dbinom{r+2}{r} + \cdots + \dbinom{r+k-1}{r} + \dbinom{r+k+1-1}{r} = \dbinom{r+k+1}{r+1}$.

$\text{LHS} = \dbinom{r+k}{r+1} + \dbinom{r+k+1-1}{r}$ by [*] ✓

$= \dbinom{r+k}{r+1} + \dbinom{r+k}{r}$

$= \dbinom{r+k+1}{r+1}$ (from part **i**)

$= \text{RHS}$

$\therefore P(k+1)$ is true.

So by mathematical induction, $P(n)$ is true for all integers $n \geq 1$. ✓

> This is a straightforward induction series proof using the solution for part **i**.

HSC exam topic grid (2011–2020)

This table shows the coverage of this topic in past HSC exams by question number. The past exams can be downloaded from the NESA website (www.educationstandards.nsw.edu.au) by selecting 'Year 11 – Year 12', 'HSC exam papers'. NESA marking feedback and guidelines can also be found there.

The new Mathematics Extension 1 course was first examined in 2020. For exams before 2020, select 'Year 11 – Year 12', 'Resources archive', 'HSC exam papers archive'.

	Series proofs	Divisibility proofs
2011	6(a)	
2012		12(a)
2013	14(a)	
2014		13(a)
2015	13(c)	
2016	14(a)	
2017		14(a)
2018	13(a)	
2019	14(a)	
2020 new course	12(a)	

CHAPTER 2 TOPIC EXAM

Vectors

ME-V1 Introduction to vectors

V1.1 Introduction to vectors

V1.2 Further operations with vectors

V1.3 Projectile motion

- A reference sheet is provided on page 155 at the back of this book.
- For questions in Section II, show relevant mathematical reasoning and/or calculations.

Reading time: 5 minutes
Working time: 1 hour
Total marks: 35

Section I – 5 questions, 5 marks
- Attempt Questions 1–5
- Allow about 8 minutes for this section

Section II – 2 questions, 30 marks
- Attempt Questions 6–7
- Allow about 52 minutes for this section

Section I

> - Attempt Questions 1–5
> - Allow about 8 minutes for this section
>
> **5 marks**

Question 1

The vectors $\underset{\sim}{u}$, $\underset{\sim}{v}$ and $\underset{\sim}{w}$ are shown.

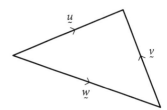

Which of the following equations is NOT true?

A $\underset{\sim}{u} - \underset{\sim}{v} = \underset{\sim}{w}$ **B** $\underset{\sim}{u} - \underset{\sim}{w} = \underset{\sim}{v}$ **C** $\underset{\sim}{u} + \underset{\sim}{w} = \underset{\sim}{v}$ **D** $\underset{\sim}{v} + \underset{\sim}{w} = \underset{\sim}{u}$

Question 2

Consider the vectors $\underset{\sim}{a} = \begin{pmatrix} 1 \\ 2 \end{pmatrix}$, $\underset{\sim}{b} = \begin{pmatrix} 2 \\ 3 \end{pmatrix}$ and $\underset{\sim}{c} = \begin{pmatrix} -1 \\ 0 \end{pmatrix}$.

Find the values of p and q such that $p\underset{\sim}{a} + q\underset{\sim}{b} = \underset{\sim}{c}$.

A $p = -3$ and $q = 1$ **B** $p = -3$ and $q = 2$

C $p = 3$ and $q = -1$ **D** $p = 3$ and $q = -2$

Question 3

What is the angle between the vectors $2\underset{\sim}{i} - 6\underset{\sim}{j}$ and $\underset{\sim}{i} + 3\underset{\sim}{j}$?

A $\cos^{-1}\left(-\dfrac{4}{5}\right)$ **B** $\cos^{-1}\left(-\dfrac{1}{25}\right)$

C $\cos^{-1}\left(\dfrac{1}{25}\right)$ **D** $\cos^{-1}\left(\dfrac{4}{5}\right)$

Question 4

Which vector is perpendicular to $\begin{pmatrix} 2 \\ 3 \end{pmatrix}$?

A $\begin{pmatrix} -3 \\ 2 \end{pmatrix}$ **B** $\begin{pmatrix} -2 \\ 3 \end{pmatrix}$ **C** $\begin{pmatrix} 2 \\ -3 \end{pmatrix}$ **D** $\begin{pmatrix} 3 \\ 2 \end{pmatrix}$

Question 5

What is the length of the projection of $\underset{\sim}{i} + \underset{\sim}{j}$ onto $2\underset{\sim}{i} - \underset{\sim}{j}$?

A $\dfrac{1}{5}$ **B** $\dfrac{\sqrt{5}}{5}$ **C** 1 **D** $\sqrt{5}$

Section II

• Attempt Questions 6–7	**30 marks**
• Allow about 52 minutes for this section	
• Answer the questions in the spaces provided. These spaces provide guidance for the expected length of response.	
• Your responses should include relevant mathematical reasoning and/or calculations.	

Question 6 (15 marks)

a   If $\underset{\sim}{i} + (k + 1)\underset{\sim}{j}$ is perpendicular to $2\underset{\sim}{i} - k\underset{\sim}{j}$, what are the values of k?   3 marks

b   Find the 2 unit vectors that are at an angle of 60° to $\begin{pmatrix} 1 \\ 1 \end{pmatrix}$.   3 marks c The sides of a parallelogram are the vectors $3\underset{\sim}{i} - \underset{\sim}{j}$ and $\underset{\sim}{i} + 2\underset{\sim}{j}$.   2 marks

Find in vector form the diagonals of the parallelogram.

d The diagram below shows the rhombus $OPQR$.

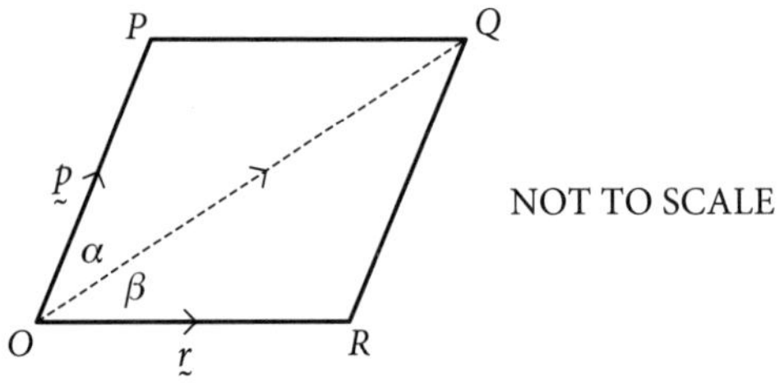

NOT TO SCALE

Let $\underset{\sim}{p} = \overrightarrow{OP}$ and $\underset{\sim}{r} = \overrightarrow{OR}$, where $\left|\underset{\sim}{p}\right| = \left|\underset{\sim}{r}\right|$. Let $\angle POQ = \alpha$ and $\angle ROQ = \beta$.

i Show that $\cos\alpha = \dfrac{\underset{\sim}{p} \cdot (\underset{\sim}{p} + \underset{\sim}{r})}{\left|\underset{\sim}{p}\right|\left|\underset{\sim}{p} + \underset{\sim}{r}\right|}$.   3 marks

ii Hence, prove that $\alpha = \beta$.   3 marks

iii What property of a rhombus is proven by the result in part **ii**?   1 mark

Question 7 (15 marks)

a Two forces, F_1 and F_2, are applied to an object in the plane. F_1 has magnitude 50 N in the   4 marks

direction $\begin{pmatrix} 3 \\ 4 \end{pmatrix}$ and F_2 has magnitude 65 N in the direction $\begin{pmatrix} 12 \\ -5 \end{pmatrix}$.

Find the vector for the total force applied to the object, and calculate its magnitude correct to the nearest N.

b An object is projected from the origin over level ground, with velocity u m/s at an angle of θ to the origin. The equation of motion for the object is $\underset{\sim}{a} = \begin{pmatrix} 0 \\ -g \end{pmatrix}$, where g m/s^2 is the acceleration due to gravity.

i Show that the position vector of the object at time t seconds is given by   3 marks

$$\underset{\sim}{r} = \begin{pmatrix} ut \cos \theta \\ ut \sin \theta - \frac{1}{2}gt^2 \end{pmatrix}.$$

ii Show that the range of the object in metres is $\dfrac{2u^2}{g} \sin \theta \cos \theta$.   2 marks

iii Hence, show that the speed of the object when it returns to the ground is independent of θ.   2 marks c   The diagram below shows ΔOAB where $\overrightarrow{OA} = \underset{\sim}{a}$ and $\overrightarrow{OB} = \underset{\sim}{b}$.

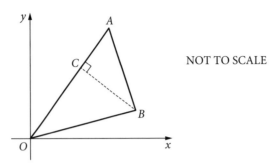

NOT TO SCALE

C lies on OA such that CB is perpendicular to OA.

i   Show that $\overrightarrow{BC} = \dfrac{1}{\underset{\sim}{a} \cdot \underset{\sim}{a}}\left[(\underset{\sim}{a} \cdot \underset{\sim}{b})\underset{\sim}{a} - (\underset{\sim}{a} \cdot \underset{\sim}{a})\underset{\sim}{b}\right].$ 3 marks

ii Hence, show that the area of ΔOAB is $\dfrac{1}{2|\underset{\sim}{a}|}\left|(\underset{\sim}{a} \cdot \underset{\sim}{b})\underset{\sim}{a} - (\underset{\sim}{a} \cdot \underset{\sim}{a})\underset{\sim}{b}\right|.$ 1 mark

END OF PAPER

WORKED SOLUTIONS

Section I (1 mark each)

Question 1

C From the diagram, $u + w = v$.

A, **B** and **D** are variations of this.

Question 2

D $p + 2q = -1$ [1]

$2p + 3q = 0$ [2]

Multiplying [1] by 2:

$2p + 4q = -2$ [3]

[3] − [2]:

$q = -2$

Substitute into [1]:

$p + 2(-2) = -1$

$p - 4 = -1$

$p = 3$

So $p = 3$, $q = -2$.

Many questions involving vectors are made easier by considering the horizontal and vertical components separately.

Question 3

A $u \cdot v = |u||v|\cos\theta$ (from the reference sheet)

Let $u = 2i - 6j$, $v = i + 3j$

$u \cdot v = 2 \times 1 + (-6) \times 3$

$\qquad = -16$

$|u| = \sqrt{2^2 + (-6)^2}$

$\qquad = \sqrt{40}$

$|v| = \sqrt{1^2 + 3^2}$

$\qquad = \sqrt{10}$

$\therefore -16 = \sqrt{40}\sqrt{10}\cos\theta$

$\qquad\quad = \sqrt{400}\cos\theta$

$\qquad\quad = 20\cos\theta$

$-\dfrac{4}{5} = \cos\theta$

$\theta = \cos^{-1}\left(-\dfrac{4}{5}\right)$

This is a straightforward question involving the scalar product of 2 vectors − use the HSC exam reference sheet.

Question 4

A Perpendicular vectors have a scalar product of 0.

$u \cdot v = 0$.

By trial and error:

$$\begin{pmatrix} 2 \\ 3 \end{pmatrix} \cdot \begin{pmatrix} -3 \\ 2 \end{pmatrix} = -6 + 6 = 0$$

This is a straightforward trial-and-error question.

Note: All vectors perpendicular to $\begin{pmatrix} x \\ y \end{pmatrix}$ are proportional to $\begin{pmatrix} -y \\ x \end{pmatrix}$.

Question 5

B The length of the projection of v onto u is

$\left|\text{proj}_u v\right| = \dfrac{u \cdot v}{|u|}$ (or $v \cdot \hat{u}$).

Let $u = 2i - j$, $v = i + j$

$u \cdot v = 2 \times 1 + (-1) \times 1$

$\qquad = 1$

$|u| = \sqrt{2^2 + (-1)^2}$

$\qquad = \sqrt{5}$

$\therefore \left|\text{proj}_u v\right| = \dfrac{1}{\sqrt{5}}$

It is much easier to understand the idea behind the projections than to find the projection and then the length.

Section II ($\checkmark$ = 1 mark)

Question 6 (15 marks)

a Perpendicular vectors have a scalar product of 0; that is, $\underset{\sim}{u} \cdot \underset{\sim}{v} = 0$.

$$(\underset{\sim}{i} + (k+1)\underset{\sim}{j}) \cdot (2\underset{\sim}{i} - k\underset{\sim}{j}) = 0 \quad \checkmark$$
$$2 - k(k+1) = 0$$
$$2 - k - k^2 = 0$$
$$k^2 + k - 2 = 0 \quad \checkmark$$
$$(k+2)(k-1) = 0$$
$$k = -2, 1 \quad \checkmark$$

This is a scalar product equation that becomes a quadratic equation.

b $\underset{\sim}{u} \cdot \underset{\sim}{v} = |\underset{\sim}{u}||\underset{\sim}{v}|\cos\theta$ (from the reference sheet)

Let $\underset{\sim}{u} = \underset{\sim}{i} + \underset{\sim}{j}, \underset{\sim}{v} = x\underset{\sim}{i} + y\underset{\sim}{j}$ where $\underset{\sim}{v}$ is a unit vector

$$\underset{\sim}{u} \cdot \underset{\sim}{v} = 1 \times x + 1 \times y$$
$$= x + y$$
$$|\underset{\sim}{u}| = \sqrt{1^2 + 1^2} = \sqrt{2}$$
$$|\underset{\sim}{v}| = \sqrt{x^2 + y^2} = 1 \quad [\star] \quad (\underset{\sim}{v} \text{ is a unit vector})$$

$$\therefore x + y = \sqrt{2}(1)\cos 60°$$
$$x + y = \frac{\sqrt{2}}{2} \quad\quad [1] \quad \checkmark$$

Also, $x^2 + y^2 = 1$ from $[\star]$ [2]

From [1]:

$$y = \frac{\sqrt{2}}{2} - x \quad \checkmark$$

Substitute into [2]:

When $x = \dfrac{\sqrt{2} + \sqrt{6}}{4}, y = \dfrac{\sqrt{2} - \sqrt{6}}{4}$ (conjugate of x)

When $x = \dfrac{\sqrt{2} - \sqrt{6}}{4}, y = \dfrac{\sqrt{2} + \sqrt{6}}{4}$.

Thus, the 2 vectors are

$$\begin{pmatrix} \dfrac{\sqrt{2} + \sqrt{6}}{4} \\ \dfrac{\sqrt{2} - \sqrt{6}}{4} \end{pmatrix} \text{ and } \begin{pmatrix} \dfrac{\sqrt{2} - \sqrt{6}}{4} \\ \dfrac{\sqrt{2} + \sqrt{6}}{4} \end{pmatrix}. \quad \checkmark$$

This complex question will test your algebra skills, including solving simultaneous equations.

c The diagonals of a parallelogram formed by $\underset{\sim}{u}$ and $\underset{\sim}{v}$ are $\underset{\sim}{u} + \underset{\sim}{v}$ and $\underset{\sim}{u} - \underset{\sim}{v}$.

Hence, the vectors are

$$(3\underset{\sim}{i} - \underset{\sim}{j}) + (\underset{\sim}{i} + 2\underset{\sim}{j}) = 4\underset{\sim}{i} + \underset{\sim}{j} \quad \checkmark$$

and $(3\underset{\sim}{i} - \underset{\sim}{j}) - (\underset{\sim}{i} + 2\underset{\sim}{j}) = 2\underset{\sim}{i} - 3\underset{\sim}{j}.$ $\checkmark$

The vectors can also be the negative of our solutions, i.e. pointing in the opposite direction.

d i $\underset{\sim}{p} \cdot \overrightarrow{OQ} = |\underset{\sim}{p}||\overrightarrow{OQ}|\cos\alpha$ $\checkmark$

But $\overrightarrow{OQ} = \underset{\sim}{p} + \underset{\sim}{r}$ $\checkmark$

$$\therefore \underset{\sim}{p} \cdot (\underset{\sim}{p} + \underset{\sim}{r}) = |\underset{\sim}{p}||\underset{\sim}{p} + \underset{\sim}{r}|\cos\alpha$$

Hence, $\cos\alpha = \dfrac{\underset{\sim}{p} \cdot (\underset{\sim}{p} + \underset{\sim}{r})}{|\underset{\sim}{p}||\underset{\sim}{p} + \underset{\sim}{r}|}.$ $\checkmark$

This is an application of the scalar product and adding vectors.

ii Likewise, from part **i**, $\cos\beta = \dfrac{\underset{\sim}{r} \cdot (\underset{\sim}{p} + \underset{\sim}{r})}{|\underset{\sim}{r}||\underset{\sim}{p} + \underset{\sim}{r}|}.$ $\checkmark$

But $|\underset{\sim}{p}| = |\underset{\sim}{r}|$, so $\cos\beta = \dfrac{\underset{\sim}{p} \cdot (\underset{\sim}{p} + \underset{\sim}{r})}{|\underset{\sim}{p}||\underset{\sim}{p} + \underset{\sim}{r}|} = \cos\alpha.$ $\checkmark$

Therefore, since $0 \le \alpha, \beta \le \pi, \alpha = \beta.$ $\checkmark$

Part **ii** extends part **i** to apply to β using the information in the question. It is very important to read the question carefully.

iii The diagonals of a rhombus bisect its angles. $\checkmark$

Question 7 (15 marks)

a The 2 force vectors have magnitude and direction, but you need to turn the direction vectors into unit vectors to make sure the total magnitude is correct.

$$\underset{\sim}{F_1} = 50 \times \frac{1}{5}\begin{pmatrix} 3 \\ 4 \end{pmatrix} = \begin{pmatrix} 30 \\ 40 \end{pmatrix} \checkmark$$

$$\underset{\sim}{F_2} = 65 \times \frac{1}{13}\begin{pmatrix} 12 \\ -5 \end{pmatrix} = \begin{pmatrix} 60 \\ -25 \end{pmatrix} \checkmark$$

Total force $= \underset{\sim}{F_1} + \underset{\sim}{F_2} = \begin{pmatrix} 90 \\ 15 \end{pmatrix} \checkmark$

Magnitude $= \sqrt{90^2 + 15^2} = \sqrt{8325} \approx 91\,\text{N} \checkmark$

It is very important to read the question carefully.

b i $\underset{\sim}{a} = \begin{pmatrix} 0 \\ -g \end{pmatrix} \rightarrow \underset{\sim}{v} = \begin{pmatrix} 0 \\ -gt \end{pmatrix} + \underset{\sim}{c_1}$

$\underset{\sim}{v}(0) = \begin{pmatrix} u\cos\theta \\ u\sin\theta \end{pmatrix} \checkmark$

$\therefore \underset{\sim}{v} = \begin{pmatrix} u\cos\theta \\ u\sin\theta - gt \end{pmatrix} \checkmark$

$\therefore \underset{\sim}{r} = \begin{pmatrix} ut\cos\theta \\ ut\sin\theta - \frac{1}{2}gt^2 \end{pmatrix} + \underset{\sim}{c_2}$

$\underset{\sim}{r}(0) = \begin{pmatrix} 0 \\ 0 \end{pmatrix} \rightarrow \underset{\sim}{r} = \begin{pmatrix} ut\cos\theta \\ ut\sin\theta - \frac{1}{2}gt^2 \end{pmatrix} \checkmark$

Projectile motion has always been in the Mathematics Extension 1 course, but the vector form of this motion was only introduced in 2020. Learn how to derive these equations of motion well because this is a FAQ in HSC exams (2020 Mathematics Extension 2 HSC exam, Question **12(b)(i)**).

ii $x = $ the range when $y = 0$.

$\therefore ut\sin\theta - \frac{1}{2}gt^2 = 0$

$\rightarrow u\sin\theta - \frac{1}{2}gt = 0 \quad (t \neq 0)$

Hence, $t = \frac{2u}{g}\sin\theta. \checkmark$

Therefore, the range

$= u\left(\frac{2u}{g}\sin\theta\right)\cos\theta$

$= \frac{2u^2}{g}\sin\theta\cos\theta \checkmark$

This is a common type of question. Note the link to part **i**. Also note that the answer can also be written as $\frac{u^2}{g}\sin 2\theta$.

iii When the object returns to the ground:

$\underset{\sim}{v} = \begin{pmatrix} u\cos\theta \\ u\sin\theta - gt \end{pmatrix}$

$= \begin{pmatrix} u\cos\theta \\ u\sin\theta - g\left(\frac{2u}{g}\sin\theta\right) \end{pmatrix}$

(from parts **i** and **ii**)

$= \begin{pmatrix} u\cos\theta \\ u\sin\theta - 2u\sin\theta \end{pmatrix}$

$= \begin{pmatrix} u\cos\theta \\ -u\sin\theta \end{pmatrix}. \checkmark$

Hence, the speed

$|\underset{\sim}{v}| = \sqrt{(u\cos\theta)^2 + (-u\sin\theta)^2}$

$= \sqrt{u^2\cos^2\theta + u^2\sin^2\theta}$

$= \sqrt{u^2(\cos^2\theta + \sin^2\theta)}$

$= \sqrt{u^2(1)}$

$= u$, which is independent of $\theta. \checkmark$

Note the links to parts **i** and **ii**. The speed of an object is the magnitude of its velocity vector. And our result indicates that only the starting speed u affects the landing speed regardless of angle.

c i $\overrightarrow{OC}$ is the projection of $\overrightarrow{OB}$ onto $\overrightarrow{OA}$. ✓

$$\therefore \overrightarrow{OC} = \text{proj}_{\underset{\sim}{a}}\underset{\sim}{b} = \frac{\underset{\sim}{b} \cdot \underset{\sim}{a}}{\underset{\sim}{a} \cdot \underset{\sim}{a}}\underset{\sim}{a}$$

$$\overrightarrow{BC} = \overrightarrow{OC} - \overrightarrow{OB} \checkmark$$

$$= \frac{\underset{\sim}{b} \cdot \underset{\sim}{a}}{\underset{\sim}{a} \cdot \underset{\sim}{a}}\underset{\sim}{a} - \underset{\sim}{b}$$

$$= \frac{\underset{\sim}{a} \cdot \underset{\sim}{b}}{\underset{\sim}{a} \cdot \underset{\sim}{a}}\underset{\sim}{a} - \frac{\underset{\sim}{a} \cdot \underset{\sim}{a}}{\underset{\sim}{a} \cdot \underset{\sim}{a}}\underset{\sim}{b}$$

$$= \frac{1}{\underset{\sim}{a} \cdot \underset{\sim}{a}}\left[(\underset{\sim}{a} \cdot \underset{\sim}{b})\underset{\sim}{a} - (\underset{\sim}{a} \cdot \underset{\sim}{a})\underset{\sim}{b}\right] \checkmark$$

When answering questions involving perpendicular vectors, you should look for possible applications of the projection.

ii Area of ΔOAB

$$= \frac{1}{2}|OA||BC|$$

$$= \frac{1}{2}|\underset{\sim}{a}|\left|\frac{1}{\underset{\sim}{a} \cdot \underset{\sim}{a}}\left[(\underset{\sim}{a} \cdot \underset{\sim}{b})\underset{\sim}{a} - (\underset{\sim}{a} \cdot \underset{\sim}{a})\underset{\sim}{b}\right]\right|$$

$$= \frac{1}{2}|\underset{\sim}{a}|\left|\frac{1}{|\underset{\sim}{a}|^2}\left[(\underset{\sim}{a} \cdot \underset{\sim}{b})\underset{\sim}{a} - (\underset{\sim}{a} \cdot \underset{\sim}{a})\underset{\sim}{b}\right]\right|$$

$$= \frac{1}{2|\underset{\sim}{a}|}\left|(\underset{\sim}{a} \cdot \underset{\sim}{b})\underset{\sim}{a} - (\underset{\sim}{a} \cdot \underset{\sim}{a})\underset{\sim}{b}\right| \checkmark$$

'Hence' usually indicates that a result in the previous part should be used to answer this part of the question.

HSC exam topic grid (2011–2020)

This table shows the coverage of this topic in past HSC exams by question number. The past exams can be downloaded from the NESA website (www.educationstandards.nsw.edu.au) by selecting 'Year 11 – Year 12', 'HSC exam papers'. NESA marking feedback and guidelines can also be found there.

The new Mathematics Extension 1 course was first examined in 2020. For exams before 2020, select 'Year 11 – Year 12', 'Resources archive', 'HSC exam papers archive'.

Vectors were introduced to the Mathematics Extension 1 course in 2020.

	Operations with vectors	Applying vectors	Projectile motion
2011			6(b)
2012			14(b)
2013			13(c)
2014			14(a)
2015			14(a)
2016			13(b)
2017			13(c)
2018			13(c)
2019			13(d)
2020 new course	6	4, 9, 11(b)	

CHAPTER 3
TOPIC EXAM

3

Trigonometric equations

ME-T3 Trigonometric equations

• A reference sheet is provided on page 155 at the back of this book • For questions in Section II, show relevant mathematical reasoning and/or calculations.	**Reading time: 5 minutes** **Working time: 1 hour** **Total marks: 35**

Section I – 5 questions, 5 marks
• Attempt Questions 1–5
• Allow about 8 minutes for this section

Section II – 2 questions, 30 marks
• Attempt Questions 6–7
• Allow about 52 minutes for this section

Section I

• Attempt Questions 1–5 **5 marks** • Allow about 8 minutes for this section

Question 1

Which expression is equal to $\sin x + \cos x$?

A $\dfrac{1}{\sqrt{2}}\sin\left(x + \dfrac{\pi}{4}\right)$

B $\dfrac{1}{\sqrt{2}}\cos\left(x + \dfrac{\pi}{4}\right)$

C $\sqrt{2}\sin\left(x + \dfrac{\pi}{4}\right)$

D $\sqrt{2}\cos\left(x + \dfrac{\pi}{4}\right)$

Question 2

What is the value of $\tan\alpha$ when the expression $2\cos x - 3\sin x$ is written in the form $\sqrt{13}\cos(x + \alpha)$?

A $-\dfrac{3}{2}$

B $-\dfrac{2}{3}$

C $\dfrac{2}{3}$

D $\dfrac{3}{2}$

Question 3

If $\tan A = 2$, what is the value of $\cos 2A$?

A $-\dfrac{3}{5}$

B $\dfrac{1}{5}$

C $\dfrac{1}{2}$

D $\dfrac{4}{5}$

Question 4

What is the general solution to the equation $2\cos^2 x - 3\cos x - 2 = 0$, where n is an integer?

A $x = \dfrac{5\pi}{6} + 2n\pi, \dfrac{7\pi}{6} + 2n\pi$

B $x = \dfrac{\pi}{6} + 2n\pi, \dfrac{11\pi}{6} + 2n\pi$

C $x = \dfrac{2\pi}{3} + 2n\pi, \dfrac{4\pi}{3} + 2n\pi$

D $x = \dfrac{\pi}{3} + 2n\pi, \dfrac{5\pi}{3} + 2n\pi$

Question 5

If $\sin\theta = \dfrac{2}{3}$, where $\dfrac{\pi}{2} < \theta < \pi$, what is the value of $\sin 2\theta$?

A $-\dfrac{4\sqrt{5}}{9}$

B $-\dfrac{1}{9}$

C $\dfrac{1}{9}$

D $\dfrac{4\sqrt{5}}{9}$

Section II

> - Attempt Questions 6–7   **30 marks**
> - Allow about 52 minutes for this section
> - Answer the questions in the spaces provided. These spaces provide
>   guidance for the expected length of response.
> - Your responses should include relevant mathematical reasoning
>   and/or calculations.

Question 6 (15 marks)

a **i** Write $\sqrt{5}\sin x + 2\cos x$ in the form $R\sin(x + \alpha)$, where $0 < \alpha < \dfrac{\pi}{2}$. Leave α in exact form.   2 marks

ii Hence, solve $\sqrt{5}\sin x + 2\cos x = 1$ for $0 \le x \le 2\pi$, correct to three decimal places.   2 marks

b Solve the equation $\sin^2\theta = 2 + \sin\theta$ for $0 \le \theta \le 2\pi$.   2 marks

c i Prove that $\cos 3x = 4\cos^3 x - 3\cos x$.   2 marks

ii Hence, solve the equation $2\cos x = \sqrt{\cos 3x}$ for $0 \le x \le 2\pi$.   3 marks

d By expressing $2\cos 2\theta + 2\sin 2\theta$ in the form $R\cos(2\theta - \alpha)$, solve $2\cos 2\theta + 2\sin 2\theta = \sqrt{6}$ 4 marks
for $0 \le \theta \le 2\pi$.

Question 7 (15 marks)

a **i** Show that $2\cos 3\theta \cos 2\theta = \cos 5\theta + \cos \theta$.   1 mark

ii Hence, solve the equation $\cos \theta + \cos 3\theta + \cos 5\theta = 0$ for $0 \le x \le 2\pi$.   3 marks

b **i** Prove that $\cos x = \sqrt{\dfrac{1 + \cos 2x}{2}}$ for $0 \le x \le \dfrac{\pi}{2}$.   1 mark

ii Hence, find the exact value of $\cos \dfrac{\pi}{24}$.   3 marks

c i Show that $\cos^4 \alpha - \cos^2 \alpha = \frac{1}{8}(\cos 4\alpha - 1)$.   2 marks

ii By letting $x = 2\cos\alpha$ in the quartic equation $x^4 - 4x^2 + 2 = 0$, show that $\cos 4\alpha = 0$.   2 marks

iii Hence, prove that $\cos^2 \frac{\pi}{8} + \cos^2 \frac{3\pi}{8} = 1$.   3 marks

END OF PAPER

WORKED SOLUTIONS

Section I (1 mark each)

Question 1

C Let $\sin x + \cos x = R \sin(x + \alpha)$
$$= R(\sin x \cos \alpha + \cos x \sin \alpha)$$
$$= R \sin x \cos \alpha + R \cos x \sin \alpha$$

$\therefore R \cos \alpha = 1 \quad [1]$

$\therefore R \sin \alpha = 1 \quad [2]$

$[2] \div [1]: \qquad \tan \alpha = \dfrac{1}{1} = 1$

$$\alpha = \frac{\pi}{4}$$

$[1]^2 + [2]^2: \quad R^2 \cos^2 \alpha + R^2 \sin^2 \alpha = 1^2 + 1^2$
$$R^2(\cos^2 \alpha + \sin^2 \alpha) = 2$$
$$R = \sqrt{2}$$

(OR from [1]: $\quad R \cos\left(\dfrac{\pi}{4}\right) = 1$

$$R \frac{1}{\sqrt{2}} = 1$$
$$R = \sqrt{2})$$

$\therefore \cos x + \sin x = \sqrt{2} \sin\left(x + \dfrac{\pi}{4}\right)$

Question 2

D $2\cos x - 3 \sin x = R \cos(x + \alpha)$
$$= R(\cos x \cos \alpha - \sin x \sin \alpha)$$
$$= R \cos x \cos \alpha - R \sin x \sin \alpha$$

$\therefore R \cos \alpha = 2 \quad [1]$

$\therefore R \sin \alpha = 3 \quad [2]$

$[2] \div [1]: \qquad \tan \alpha = \dfrac{3}{2}$

Questions 1 and 2 are standard auxiliary angle questions.

Question 3

A $\cos 2A = \dfrac{1 - t^2}{1 + t^2}$
$$= \frac{1 - 2^2}{1 + 2^2}$$
$$= -\frac{3}{5}$$

Make sure you use the HSC exam reference sheet for this application of the *t*-formulas.

Question 4

C $2\cos^2 x - 3\cos x - 2 = (2\cos x + 1)(\cos x - 2)$

$\therefore \cos x = -\dfrac{1}{2}$

$x = \dfrac{2\pi}{3}, \dfrac{4\pi}{3}, \dfrac{8\pi}{3}, \ldots$

$x = \dfrac{2\pi}{3} + 2n\pi, \dfrac{4\pi}{3} + 2n\pi$, where n is an integer.

Question 5

A $\cos \theta = -\sqrt{1 - \sin^2 \theta}$ when $\dfrac{\pi}{2} < \theta < \pi$

$$= -\sqrt{1 - \left(\frac{2}{3}\right)^2}$$
$$= -\sqrt{1 - \frac{4}{9}}$$
$$= -\sqrt{\frac{5}{9}}$$
$$= -\frac{\sqrt{5}}{3}$$

$\therefore \sin 2\theta = 2 \sin \theta \cos \theta$
$$= 2 \times \frac{2}{3} \times \left(-\frac{\sqrt{5}}{3}\right)$$
$$= -\frac{4\sqrt{5}}{9}$$

Make sure you identify the sign of the trigonometric ratios. $\cos \theta$ is negative in the 2nd quadrant.

Section II ($\checkmark$ = 1 mark)

Question 6 (15 marks)

a i Let $\sqrt{5}\sin x + 2\cos x = R\sin(x + \alpha)$
$$= R(\sin x \cos \alpha + \cos x \sin \alpha)$$
$$= R\sin x \cos \alpha + R\cos x \sin \alpha$$

$\therefore R\cos \alpha = \sqrt{5}$ [1]

$\therefore R\sin \alpha = 2$ [2]

[2] ÷ [1]: $\tan \alpha = \dfrac{2}{\sqrt{5}}$

$\alpha = \tan^{-1}\left(\dfrac{2}{\sqrt{5}}\right)$ $\checkmark$

[1]^2 + [2]^2: $R^2\cos^2\alpha + R^2\sin^2\alpha = (\sqrt{5})^2 + 2^2$
$$R^2(\cos^2\alpha + \sin^2\alpha) = 9$$
$$R = 3$$

$\therefore \sqrt{5}\sin x + 2\cos x = 3\sin\left(x + \tan^{-1}\dfrac{2}{\sqrt{5}}\right)$ $\checkmark$

This is an auxiliary angle method question.

ii $3\sin\left(x + \tan^{-1}\left(\dfrac{2}{\sqrt{5}}\right)\right) = 1$

$\sin\left(x + \tan^{-1}\left(\dfrac{2}{\sqrt{5}}\right)\right) = \dfrac{1}{3}$ $\checkmark$

$x + \tan^{-1}\left(\dfrac{2}{\sqrt{5}}\right) = 0.33983\ldots$

$x + 0.72972\ldots = 0.33983\ldots$

$x = -0.38989\ldots$

But $0 \le x \le 2\pi$,

so $0.72972\ldots \le x + 0.72972\ldots \le 2\pi + 0.72972\ldots$
$$0.72972\ldots \le x + 0.72972\ldots \le 7.01290\ldots$$

$\therefore x + 0.72972\ldots = \pi - 0.33983\ldots$ or $2\pi + 0.33983\ldots$
          (2nd quadrant)   (1st quadrant)
$x + 0.72972\ldots = 2.80175\ldots$ or $6.62301\ldots$
$$x \approx 2.072, 5.893 \ \checkmark$$

Be wary of the domain of $\left(x + \tan^{-1}\left(\dfrac{2}{\sqrt{5}}\right)\right)$; always check that your answers are within the bounds.

If the question is in radians, make sure to answer in radians.

b $\sin^2\theta = 2 + \sin\theta$
$$\sin^2\theta - \sin\theta - 2 = 0$$
$$(\sin\theta - 2)(\sin\theta + 1) = 0 \ \checkmark$$
$$\sin\theta = -1$$
$$\theta = \dfrac{3\pi}{2} \ \checkmark$$

Using a substitution such as $a = \sin\theta$ can help in recognising the quadratic.

c **i** LHS $= \cos 3x$

$\qquad = \cos(2x + x)$

$\qquad = \cos 2x \cos x - \sin 2x \sin x$

$\qquad = (\cos^2 x - \sin^2 x)(\cos x) - 2\sin x \cos x \sin x$ ✓

$\qquad = (2\cos^2 x - 1)(\cos x) - 2\sin^2 x \cos x$

$\qquad = 2\cos^3 x - \cos x - 2(1 - \cos^2 x)\cos x$

$\qquad = 2\cos^3 x - \cos x - 2\cos x + 2\cos^3 x$

$\qquad = 4\cos^3 x - 3\cos x$

$\qquad = \text{RHS}$ ✓

ii $2\cos x = \sqrt{\cos 3x}$

$4\cos^2 x = \cos 3x \, (\cos x \ge 0)$

$\therefore 4\cos^2 x = 4\cos^3 x - 3\cos x$ (from part **i**) ✓

$\qquad 4\cos^3 x - 4\cos^2 x - 3\cos x = 0$

$\qquad \cos x(4\cos^2 x - 4\cos x - 3) = 0$

$\quad \cos x(2\cos x - 3)(2\cos x + 1) = 0$ ✓

$\cos x = 0, \dfrac{3}{2}, -\dfrac{1}{2}$

$\cos x = \dfrac{3}{2}$ has no solution and $\cos x \ge 0$

$\cos x = 0$

$\qquad x = \dfrac{\pi}{2}, \dfrac{3\pi}{2}$ ✓

Be careful when squaring equations that you don't double the number of solutions.

d Let $2\sin 2\theta + 2\cos 2\theta = R\cos(2\theta - \alpha)$

$\qquad\qquad\qquad\qquad\quad = R(\cos 2\theta \cos \alpha + \sin 2\theta \sin \alpha)$

$\qquad\qquad\qquad\qquad\quad = R\cos 2\theta \cos \alpha + R\sin 2\theta \sin \alpha)$

$\therefore R\cos\alpha = 2$ [1]

$\therefore R\sin\alpha = 2$ [2]

[2] ÷ [1]: $\tan\alpha = \dfrac{2}{2} = 1$

$\qquad\qquad\qquad \alpha = \dfrac{\pi}{4}$ ✓

$[1]^2 + [2]^2$: $R^2\cos^2\alpha + R^2\sin^2\alpha = 2^2 + 2^2$

$\qquad\qquad\qquad R^2(\cos^2\alpha + \sin^2\alpha) = 8$

$\qquad\qquad\qquad\qquad\qquad\qquad R = \sqrt{8}$

$\qquad\qquad\qquad\qquad\qquad\qquad\quad = 2\sqrt{2}$

(OR from [1]: $R\cos\left(\dfrac{\pi}{4}\right) = 2$

$\qquad\qquad\qquad\quad R\dfrac{1}{\sqrt{2}} = 2$

$\qquad\qquad\qquad\quad\quad R = 2\sqrt{2}$)

Hence, $2\cos 2\theta + 2\sin 2\theta = 2\sqrt{2}\cos\left(2\theta - \dfrac{\pi}{4}\right)$ ✓

$2\cos 2\theta + 2\sin 2\theta = 2\sqrt{2}\cos\left(2\theta - \dfrac{\pi}{4}\right) = \sqrt{6}$

Hence, $\cos\left(2\theta - \dfrac{\pi}{4}\right) = \dfrac{\sqrt{6}}{2\sqrt{2}} = \dfrac{\sqrt{3}}{2}$ ✓

Thus, $2\theta - \dfrac{\pi}{4} = \dfrac{\pi}{6}, \dfrac{\pi}{6}, \dfrac{11\pi}{6}, \dfrac{13\pi}{6}$

$2\theta = \dfrac{\pi}{12}, \dfrac{5\pi}{12}, \dfrac{25\pi}{12}, \dfrac{29\pi}{12}$

$\theta = \dfrac{\pi}{24}, \dfrac{5\pi}{24}, \dfrac{25\pi}{24}, \dfrac{29\pi}{24}$ ✓

Note that this type of question does not need to be broken into 2 parts. Also note there are 4 solutions because $0 \le 2\theta \le 4\pi$.

Question 7 (15 marks)

a i   LHS $= 2\cos 3\theta \cos 2\theta$

$= 2 \times \dfrac{1}{2}[\cos(3\theta + 2\theta) + \cos(3\theta - 2\theta)]$ (from the reference sheet)

$= \cos 5\theta + \cos \theta$

$= $ RHS ✓

Make sure that in 'show that' questions you show enough working, or the formula you based it on.

ii $\cos\theta + \cos 3\theta + \cos 5\theta = 0$

$\cos 3\theta + 2\cos 3\theta \cos 2\theta = 0$ (from part **i**) ✓

$\cos 3\theta(1 + 2\cos 2\theta) = 0$

$\cos 3\theta = 0$ or $2\cos 2\theta = -1$

$\cos 3\theta = 0$ or $\cos 2\theta = -\dfrac{1}{2}$ ✓

Hence, $3\theta = \dfrac{\pi}{2}, \dfrac{3\pi}{2}, \dfrac{5\pi}{2}, \dfrac{7\pi}{2}, \dfrac{9\pi}{2}, \dfrac{11\pi}{2}$ or $2\theta = \dfrac{2\pi}{3}, \dfrac{4\pi}{3}, \dfrac{8\pi}{3}, \dfrac{10\pi}{3}$.

So $\theta = \dfrac{\pi}{6}, \dfrac{\pi}{3}, \dfrac{\pi}{2}, \dfrac{2\pi}{3}, \dfrac{5\pi}{6}, \dfrac{7\pi}{6}, \dfrac{4\pi}{3}, \dfrac{3\pi}{2}, \dfrac{5\pi}{3}, \dfrac{11\pi}{6}$. ✓

Note that there are $6 + 4 = 10$ solutions in total (corresponding to $\dfrac{k\pi}{6}$ for $k = 1, 2, 3, 4, 5, 6, 7, 8, 10, 11$). Always remember to use the solution from part **i** in part **ii**.

b i $\cos^2 x = \dfrac{1}{2}(1 + \cos 2x)$ (from the reference sheet)

Hence, $\cos x = \sqrt{\dfrac{1 + \cos 2x}{2}}$ as $\cos x \ge 0$ in $0 \le x \le \dfrac{\pi}{2}$. ✓

Make sure you have at least one line of working. Take note of the correct domain and only include the positive root for $\cos x$.

ii $\cos\dfrac{\pi}{24} = \sqrt{\dfrac{1 + \cos\dfrac{\pi}{12}}{2}}$ (from part **i**)

$\cos\dfrac{\pi}{12} = \sqrt{\dfrac{1 + \cos\dfrac{\pi}{6}}{2}}$ ✓ (from part **i**)

$= \sqrt{\dfrac{1 + \dfrac{\sqrt{3}}{2}}{2}}$

$= \sqrt{\dfrac{2 + \sqrt{3}}{4}}$

$= \dfrac{1}{2}\sqrt{2 + \sqrt{3}}$ ✓

WORKED SOLUTIONS

$$\text{So } \cos\frac{\pi}{24} = \sqrt{\frac{1 + \frac{1}{2}\sqrt{2+\sqrt{3}}}{2}}$$

$$= \sqrt{\frac{2+\sqrt{2+\sqrt{3}}}{4}}$$

$$= \frac{1}{2}\sqrt{2+\sqrt{2+\sqrt{3}}}. \checkmark$$

Make sure you simplify when possible to avoid making mistakes, and always check your answers.
Note: the triple surd can be simplified, but it is not necessary.

c i $\cos 4\alpha = 2\cos^2 2\alpha - 1$

$$= 2(2\cos^2\alpha - 1)^2 - 1$$

$$= 2(4\cos^4\alpha - 4\cos^2\alpha + 1) - 1$$

$$= 8\cos^4\alpha - 8\cos^2\alpha + 1 \checkmark$$

$$\text{LHS} = \cos^4\alpha - \cos^2\alpha$$

$$= \frac{1}{8}(\cos 4\alpha - 1) \checkmark$$

This complex identity is a variation of Question 14(b)(i) of the 2020 HSC exam.

ii Substituting:

$$(2\cos\alpha)^4 - 4(2\cos\alpha)^2 + 2 = 0$$

$$16\cos^4\alpha - 16\cos^2\alpha + 2 = 0 \checkmark$$

$$8\cos^4\alpha - 8\cos^2\alpha + 1 = 0$$

$$8(\cos^4\alpha - \cos^2\alpha) + 1 = 0$$

Therefore, from part **i**:

$$(\cos 4\alpha - 1) + 1 = 0$$

$$\cos 4\alpha = 0 \checkmark$$

iii The solutions to $\cos 4\alpha = 0$ are $\alpha = \dfrac{\pi}{8}, \dfrac{3\pi}{8}, \dfrac{5\pi}{8}, \dfrac{7\pi}{8}, \dfrac{9\pi}{8}, \dfrac{11\pi}{8}, \dfrac{13\pi}{8}, \dfrac{15\pi}{8}$ for $0 \le \alpha \le 2\pi$.

But note that $\cos\dfrac{\pi}{8} = \cos\dfrac{7\pi}{8}$, etc.

Hence, the roots of the quartic equation are $x = 2\cos\dfrac{\pi}{8}, 2\cos\dfrac{3\pi}{8}, 2\cos\dfrac{9\pi}{8}, 2\cos\dfrac{11\pi}{8}$. $\checkmark$

Note also that $2\cos\dfrac{\pi}{8} = -2\cos\dfrac{9\pi}{8}$ and $2\cos\dfrac{3\pi}{8} = -2\cos\dfrac{11\pi}{8}$.

Thus, the roots are $\beta, -\beta, \gamma, -\gamma$.

From Viete's formula for the sum of the roots in pairs:

$$\beta(-\beta) + \beta\gamma + \beta(-\gamma) - \beta\gamma - \beta(-\gamma) + \gamma(-\gamma) = -\frac{4}{1} \checkmark$$

$$-\beta^2 - \gamma^2 = -4$$

Hence, $\beta^2 + \gamma^2 = 4$ and thus $\cos^2\dfrac{\pi}{8} + \cos^2\dfrac{3\pi}{8} = 1$. $\checkmark$

This is a very hard question that is similar to the last question of the 2020 Extension 1 HSC exam.
The key to the solution is realising that it depends on polynomial roots, and that there are only 2 distinct roots (ignoring sign).

HSC exam topic grid (2011–2020)

This table shows the coverage of this topic in past HSC exams by question number. The past exams can be downloaded from the NESA website (www.educationstandards.nsw.edu.au) by selecting 'Year 11 – Year 12', 'HSC exam papers'. NESA marking feedback and guidelines can also be found there.

The new Mathematics Extension 1 course was first examined in 2020. For exams before 2020, select 'Year 11 – Year 12', 'Resources archive', 'HSC exam papers archive'.

	Auxiliary angle method $a \cos x + b \sin x = c$	**Trigonometric equations using identities**	**Proving identities**
2011		5(a)(v)	5(a)(ii)
2012			
2013	12(a)	6	8
2014	2		
2015	11(d)		
2016		6	3
2017	4		
2018	11(c)	9	
2019	12(b)(i), (ii)		6
2020 new course	11(d)		14(b)

CHAPTER 4
TOPIC EXAM

4

Further integration

ME-C2 Further calculus skills

Note: Area and volumes of integration are covered in Topic exam 5, Volumes and differential equations.

• A reference sheet is provided on page 155 at the back of this book. • For questions in Section II, show relevant mathematical reasoning and/or calculations. **Section I – 5 questions, 5 marks** • Attempt Questions 1–5 • Allow about 8 minutes for this section **Section II – 2 questions, 30 marks** • Attempt Questions 6–7 • Allow about 52 minutes for this section	**Reading time: 5 minutes** **Working time: 1 hour** **Total marks: 35**

Section I

> - Attempt Questions 1–5   **5 marks**
> - Allow about 8 minutes for this section

Question 1

Find the anti-derivative of $\dfrac{1}{\sqrt{1-4x^2}}$.

A $\sin^{-1}x + c$

B $\sin^{-1}2x + c$

C $\dfrac{1}{2}\sin^{-1}x + c$

D $\dfrac{1}{2}\sin^{-1}2x + c$

Question 2

What is the derivative of $\cos^{-1}\left(\dfrac{x}{3}\right)$?

A $-\dfrac{1}{\sqrt{3-x^2}}$

B $-\dfrac{1}{\sqrt{9-x^2}}$

C $-\dfrac{3}{\sqrt{3-x^2}}$

D $-\dfrac{3}{\sqrt{9-x^2}}$

Question 3

Which integral is obtained when the substitution $u = \ln 2x$ is applied to $\int \dfrac{\ln 2x}{x}\,dx$?

A $\dfrac{1}{4}\int u\,du$

B $\dfrac{1}{2}\int u\,du$

C $\int u\,du$

D $2\int u\,du$

Question 4

Find $\int \cos^2\left(\dfrac{x}{2}\right)dx$.

A $\dfrac{1}{2}(x - \sin x) + c$

B $\dfrac{1}{2}(x + \sin x) + c$

C $\dfrac{1}{6}\cos^3\left(\dfrac{x}{2}\right) + c$

D $-\dfrac{1}{6}\sin^3\left(\dfrac{x}{2}\right) + c$

Question 5

Find the value of k such that $\displaystyle\int_0^k \dfrac{1}{4+x^2}\,dx = \dfrac{\pi}{6}$.

A $\dfrac{\sqrt{3}}{2}$

B 1

C $\sqrt{3}$

D $2\sqrt{3}$

Section II

TOPIC EXAM

• Attempt Questions 6–7	**30 marks**
• Allow about 52 minutes for this section	
• Answer the questions in the spaces provided. These spaces provide guidance for the expected length of response.	
• Your responses should include relevant mathematical reasoning and/or calculations.	

Question 6 (15 marks)

a Differentiate $\sin^{-1}\left(\dfrac{1}{x}\right)$. 2 marks

b Find the derivative of $x^2 \tan^{-1} x$. 2 marks

c Use the substitution $u = 1 - x^2$ to evaluate $\displaystyle\int_0^{\frac{1}{2}} \frac{x}{\sqrt{1-x^2}}\, dx$. 3 marks

d Use the substitution $x = \cos u$ to evaluate $\displaystyle\int_0^{\frac{1}{2}} \sqrt{1 - x^2}\, dx$.   4 marks

e **i** Find $\dfrac{d}{d\theta}(\tan^3 \theta)$.   1 mark

    ii Use the substitution $x = \sin \theta$ to evaluate $\displaystyle\int_0^{\frac{1}{2}} \dfrac{x^2}{(1 - x^2)^{\frac{5}{2}}}\, dx$.   3 marks

Question 7 (15 marks)

a Find $\int \dfrac{dx}{1 + 2x^2}$.   2 marks

b Use the substitution $u = 1 - x$ to evaluate $\int_{}^{} x(1 - x)^{\frac{4}{3}}\, dx$.   3 marks

c Let $f(x) = \tan^{-1}(x^2) + \tan^{-1}(x^{-2})$ for $x > 0$.

   i Show that $f'(x) = 0$ for $x > 0$.   2 marks

   ii Hence, find $\int_{1}^{3} f(x)\, dx$.   2 marks

TOPIC EXAM

d Let $f(x) = \cos(\tan^{-1}x)$ and $g(x) = (1 + x^2)^{-\frac{1}{2}}$.

   i   Show that $f'(x) = g'(x)$.   4 marks

   ii   Hence, or otherwise, show that $f(x) = g(x)$.   2 marks

END OF PAPER

WORKED SOLUTIONS

Section I (1 mark each)

Question 1

D $\dfrac{1}{\sqrt{1-4x^2}} = \dfrac{1}{2}\dfrac{2}{\sqrt{1-(2x)^2}}$

This is a common integration question involving an inverse trigonometric function – use the HSC exam reference sheet.

Question 2

B $\dfrac{d}{dx}\cos^{-1}\left(\dfrac{x}{3}\right) = -\dfrac{\frac{1}{3}}{\sqrt{1-\left(\frac{x}{3}\right)^2}}$

$\qquad\qquad\qquad = -\dfrac{1}{\sqrt{9-x^2}}$

Make sure you simplify the fractions properly.

Question 3

C $du = \dfrac{2}{2x}\,dx = \dfrac{dx}{x}$

$\displaystyle\int \dfrac{\ln 2x}{x}\,dx = \int u\,du$

This is a standard substitution question.

Question 4

B $\cos^2\left(\dfrac{x}{2}\right) = \dfrac{1}{2}(1+\cos x)$ (from the reference sheet)

$\displaystyle\int \dfrac{1}{2}(1+\cos x)\,dx = \dfrac{1}{2}(x+\sin x)+c$

This is almost a syllabus dot point question.

Question 5

D $\displaystyle\int_0^k \dfrac{1}{4+x^2}\,dx = \left[\dfrac{1}{2}\tan^{-1}\left(\dfrac{x}{2}\right)\right]_0^k$

$\qquad\qquad\qquad\qquad = \dfrac{1}{2}\tan^{-1}\left(\dfrac{k}{2}\right) - 0$

$\qquad\qquad\qquad\qquad = \dfrac{\pi}{6}$

$\tan^{-1}\left(\dfrac{k}{2}\right) = \dfrac{\pi}{3}$

$\dfrac{k}{2} = \sqrt{3}$

$k = 2\sqrt{3}$

This question requires you to be careful and to check your answers.

Section II (✓ = 1 mark)

Question 6 (15 marks)

a $\dfrac{d}{dx}\left(\dfrac{1}{x}\right) = -\dfrac{1}{x^2}$ ✓

$\dfrac{d}{dx}\sin^{-1}\left(\dfrac{1}{x}\right) = -\dfrac{\frac{1}{x^2}}{\sqrt{1-\frac{1}{x^2}}}$

$\qquad\qquad\qquad = -\dfrac{1}{x\sqrt{x^2-1}}$ ✓

b $\dfrac{d}{dx}x^2 = 2x$

$\dfrac{d}{dx}\tan^{-1}x = \dfrac{1}{1+x^2}$ ✓

$\dfrac{d}{dx}(x^2\tan^{-1}x) = 2x\tan^{-1}x + \dfrac{x^2}{1+x^2}$ ✓

Parts **a** and **b** are common types of question – use the HSC exam reference sheet.

c $u = 1-x^2$

$du = -2x\,dx$

$x\,dx = -\dfrac{1}{2}\,du$ ✓

$x = 0 \rightarrow u = 1 - 0^2 = 1$

$x = \dfrac{1}{2} \rightarrow u = 1 - \left(\dfrac{1}{2}\right)^2 = \dfrac{3}{4}$

$\displaystyle\int_0^{\frac{1}{2}} \dfrac{x}{\sqrt{1-x^2}}\,dx = \int_1^{\frac{3}{4}} \dfrac{-\frac{1}{2}}{\sqrt{u}}\,du$ ✓

$\qquad\qquad\qquad = \left[-\sqrt{u}\right]_1^{\frac{3}{4}}$

$\qquad\qquad\qquad = -\dfrac{\sqrt{3}}{2} + 1$

$\qquad\qquad\qquad = 1 - \dfrac{\sqrt{3}}{2}$ ✓

This is a common exam question – don't forget to change the limits of the integral.

d $x = \cos u$

$dx = -\sin u\, du$ ✓

$x = 0 \to u = \dfrac{\pi}{2}$

$x = \dfrac{1}{2} \to u = \dfrac{\pi}{3}$

$\displaystyle\int_0^{\frac{1}{2}} \sqrt{1 - x^2}\, dx$

$= \displaystyle\int_{\frac{\pi}{2}}^{\frac{\pi}{3}} -\sin u \sqrt{1 - \cos^2 u}\, du$ ✓

$= \displaystyle\int_{\frac{\pi}{2}}^{\frac{\pi}{3}} -\sin^2 u\, du$

$= \displaystyle\int_{\frac{\pi}{3}}^{\frac{\pi}{2}} \frac{1}{2}(1 - \cos 2u)\, du$ ✓ (from the reference sheet)

$= \left[\dfrac{1}{2}u - \dfrac{1}{4}\sin 2u\right]_{\frac{\pi}{3}}^{\frac{\pi}{2}}$

$= \dfrac{\pi}{4} - \dfrac{1}{4}\sin\pi - \dfrac{\pi}{6} + \dfrac{1}{4}\sin\dfrac{2\pi}{3}$

$= \dfrac{\pi}{12} + \dfrac{\sqrt{3}}{8}$ ✓

This question can be done geometrically, but you would not get full marks because there is no substitution.

e i $\dfrac{d}{dx}(\tan^3 x) = 3 \times \sec^2 x \times \tan^2 x$ ✓

This is a direct application of the chain rule. It may be easier to rewrite such expressions as $(\tan x)^3$.

ii $x = \sin\theta$

$dx = \cos\theta\, d\theta$ ✓

$x = 0 \to \theta = 0$

$x = \dfrac{1}{2} \to \theta = \dfrac{\pi}{6}$

$\displaystyle\int_0^{\frac{1}{2}} \frac{x^2}{(1 - x^2)^{\frac{5}{2}}}\, dx = \int_0^{\frac{\pi}{6}} \frac{\sin^2\theta\cos\theta}{(1 - \sin^2\theta)^{\frac{5}{2}}}\, d\theta$ ✓

$= \displaystyle\int_0^{\frac{\pi}{6}} \frac{\sin^2\theta\cos\theta}{\cos^5\theta}\, d\theta$

$= \displaystyle\int_0^{\frac{\pi}{6}} \tan^2\theta\sec^2\theta\, d\theta$

$= \left[\dfrac{1}{3}\tan^3\theta\right]_0^{\frac{\pi}{6}}$ (from part **i**)

$= \dfrac{1}{3}\tan^3\dfrac{\pi}{6} - \dfrac{1}{3}\tan^3 0$

$= \dfrac{1}{9\sqrt{3}}$ ✓

Even if there is a mistake in part **i**, you can still receive full marks for part **ii**.

Question 7

a $\displaystyle\int \frac{dx}{1 + 2x^2} = \int \frac{dx}{1 + (\sqrt{2}x)^2}$ ✓

$= \tan^{-1}(\sqrt{2}x)$ ✓

We can distinguish $\tan^{-1}$ integrals from the others by the lack of a square root in the denominator.

b $u = 1 - x$

$du = -dx$

$dx = -du$ ✓

$\displaystyle\int x(1 - x)^{\frac{2}{3}}\, dx = \int -(1 - u)u^{\frac{2}{3}}\, du$

$= \displaystyle\int \left(u^{\frac{5}{3}} - u^{\frac{2}{3}}\right) du$ ✓

$= \dfrac{3}{8}u^{\frac{8}{3}} - \dfrac{3}{5}u^{\frac{5}{3}} + c$

$= \dfrac{3}{8}(1 - x)^{\frac{8}{3}} - \dfrac{3}{5}(1 - x)^{\frac{5}{3}} + c$ ✓

Don't forget to substitute x back in and the '$+ c$'. Simplifying further is optional.

c i $f'(x) = \dfrac{2x}{1 + x^4} - \dfrac{2x^{-3}}{1 + x^{-4}}$ ✓

$= \dfrac{2x}{1 + x^4} - \dfrac{2x}{x^4 + 1}$

$= 0$ ✓

ii Since $f'(x) = 0$, $f(x)$ is a constant function for $x > 0$.

As $f(x)$ is constant, we can substitute a convenient value, for example, $x = 1$, to find the value of our function.

$f(1) = \tan^{-1} 1 + \tan^{-1} 1$

$= \dfrac{\pi}{4} + \dfrac{\pi}{4}$

$= \dfrac{\pi}{2}$ ✓

So $\displaystyle\int_1^3 f(x)\, dx = \int_1^3 \frac{\pi}{2}\, dx$

$= \left[\dfrac{\pi}{2}x\right]_1^3$

$= \dfrac{3\pi}{2} - \dfrac{\pi}{2}$

$= \pi$ ✓

This is a straightforward question if you realise $f(x)$ is a constant function. Use the hint from part **i** to find an easier expression for $f(x)$.

d i $f'(x) = \dfrac{1}{1+x^2}\left[-\sin(\tan^{-1}x)\right]$ ✓

Let $\tan^{-1}x = \theta$ and $-\dfrac{\pi}{2} < \theta < \dfrac{\pi}{2}$.

Then $\tan\theta = x$. So $\sin\theta = \dfrac{x}{\sqrt{1+x^2}}$

($\tan x$ and $\sin x$ have the same sign in

$-\dfrac{\pi}{2} < \theta < \dfrac{\pi}{2}$) ✓

Hence, $f'(x) = -\dfrac{x}{(1+x^2)^{\frac{3}{2}}}$.

$g'(x) = -\dfrac{1}{2}2x(1+x^2)^{-\frac{3}{2}}$ ✓

$= -\dfrac{x}{(1+x^2)^{\frac{3}{2}}}$

$= f'(x)$ ✓

This is a hard question. It depends on being able to identify that $\sin(\tan^{-1}x) = \dfrac{x}{\sqrt{1+x^2}}$.

ii Since $f(x)$ and $g(x)$ have the same derivative, they differ by only a constant. ✓

$f(0) = \cos(\tan^{-1}0)$
$= \cos 0 = 1$

$g(0) = (1+0^2)^{-\frac{1}{2}}$

$= 1^{-\frac{1}{2}} = 1 = f(0)$

Hence, $f(x) = g(x)$. ✓

This is a straightforward question. You can integrate to show that they differ by a constant.

HSC exam topic grid (2011–2020)

This grid shows the coverage of this topic in past HSC exams by question number. The past exams can be downloaded from the NESA website (www.educationstandards.nsw.edu.au) by selecting 'Year 11 – Year 12', 'HSC exam papers'. NESA marking feedback and guidelines can also be found there.

The new Mathematics Extension 1 course was first examined in 2020. Volumes of solids of revolution was in the 'Mathematics' course before 2020. For exams before 2020, select 'Year 11 – Year 12', 'Resources archive', 'HSC exam papers archive'.

	Integration by substitution	Trigonometric integrals	Inverse functions and inverse trigonometric functions	Volumes of solids of revolution
2011	1(d)			8(b)*
2012	11(d)	7	9, 11(a)	14(b)*
2013	5, 11(f)	12(b)	11(b), (g)	15(b)*, Ext 1: 12(b)
2014	11(d)	12(b)	6	14(c)*, Ext 1: 12(b)
2015	11(e)		7, 13(d)	16(b)*
2016	11(b)	5	11(c)	15(a)*
2017	11(e)	11(f)	11(b)	12(b)*, Ext 1:12(c)(i)
2018	11(f)	12(a)	12(c)	14(b)*
2019	13(a)	11(e), 13(a)	3	13(d)*
2020 new course	13(a)	12(d), 13(b)	3, 13(c)	13(b)

* Volumes of solids of revolution is covered in Topic exam 5, Volumes and differential equations. The past HSC questions listed in the table for this topic come from the Mathematics exam before 2020.

WORKED SOLUTIONS

CHAPTER 5
TOPIC EXAM

Volumes and differential equations

ME-C3 Applications of calculus

    C3.1 Further area and volumes of solids of revolution

    C3.2 Differential equations

- A reference sheet is provided on page 155 at the back of this book.
- For questions in Section II, show relevant mathematical reasoning and/or calculations.

Reading time: 5 minutes
Working time: 1 hour
Total marks: 35

Section I – 5 questions, 5 marks
- Attempt Questions 1–5
- Allow about 8 minutes for this section

Section II – 2 questions, 30 marks
- Attempt Questions 6–7
- Allow about 52 minutes for this section

9780170459259

Section I

> - Attempt Questions 1–5   **5 marks**
> - Allow about 8 minutes for this section

Question 1

Which differential equation has the slope field below?

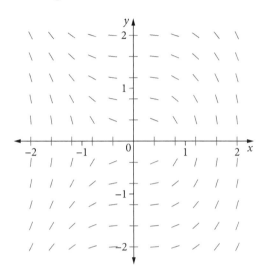

A $\dfrac{dy}{dx} = \dfrac{x^2}{y}$

B $\dfrac{dy}{dx} = \dfrac{y}{x^2}$

C $\dfrac{dy}{dx} = -\dfrac{x^2}{y}$

D $\dfrac{dy}{dx} = -\dfrac{y}{x^2}$

Question 2

The area between the curve $y = x^2$, the y-axis and the line $y = 1$ is rotated about the y-axis.

Which expression gives the volume of the solid generated?

A $\pi \int_0^1 y \, dy$

B $\pi \int_0^1 x^2 \, dx$

C $\pi \int_0^1 y^2 \, dy$

D $\pi \int_0^1 x^4 \, dx$

Question 3

Which function is a solution to the differential equation $\dfrac{dy}{dx} = 2xy$?

A $y = e^{x^2}$

B $y = e^{2x}$

C $y = 2e^x$

D $y = e^{x+2}$

Question 4

What is the direction field for the differential equation $\dfrac{dy}{dx} = y - x$?

A

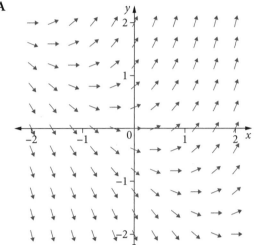

B

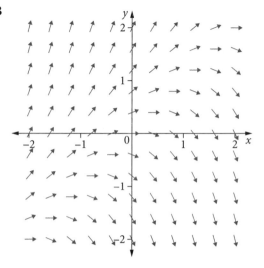

C

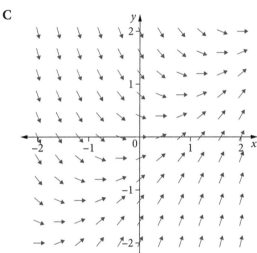

D

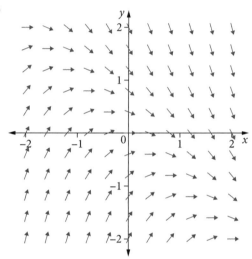

Question 5

Which differential equation has a solution $y = \sqrt{1 - x^2}$?

A $\dfrac{dy}{dx} = -\dfrac{x}{y}$

B $\dfrac{dy}{dx} = \dfrac{\cos x}{\sin y}$

C $\dfrac{dy}{dx} = \dfrac{1}{2} x^2 y^2$

D $\dfrac{dy}{dx} = x^2 + y^2$

Section II

•   Attempt Questions 6–7   **30 marks**
 •   Allow about 52 minutes for this section
 •   Answer the questions in the spaces provided. These spaces provide
     guidance for the expected length of response.
 •   Your responses should include relevant mathematical reasoning
     and/or calculations.

Question 6 (15 marks)

a   The paths of particles through the air are described by the differential equation

$$\frac{dy}{dx} = \frac{1}{6}(y+1)(y-x^2).$$

The slope field for the differential equation is sketched below.

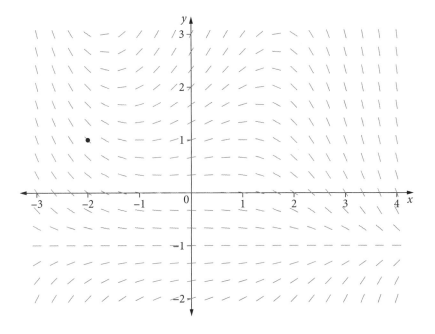

i   Identify any solutions of the form $y = c$, where c is a constant.   1 mark

_____

_____

_____

_____

ii   On the slope field, draw a sketch of the path of a particle that passes through   3 marks
     the marked point $(-2, 1)$ and describe the path as $x \to \infty$.

_____

_____

b Find the exact value of the volume of the solid of revolution formed when the region bounded by the curve $y = 2 - 2x$ and the axes is rotated about the y-axis.   3 marks

c Solve $\dfrac{dy}{dx} = \sec 2y$, finding x as a function of y.   2 marks

d Solve $\dfrac{dP}{dt} = -\dfrac{P}{t}$, where $P > 0$, $t > 0$, finding P as a function of t.   3 marks

TOPIC EXAM

e Find the function y that satisfies the differential equation $\dfrac{dy}{dx} = 2x(1 + y^2)$ and passes through the point $(0, 1)$.

3 marks

Question 7 (15 marks)

a The shaded region below is bounded by the graphs of $y = \sqrt{x}$ and $y = \dfrac{1}{2}x$.

4 marks

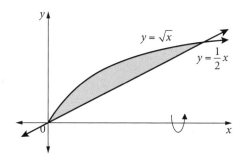

Find the volume of the solid of revolution formed when the region is rotated about the x-axis.

b The population of the world can be modelled by the logistic equation $\dfrac{dP}{dt} = \dfrac{1}{800}P(16 - P)$,

where P is the population of the world (in billion people) and t is the time in years since 2020.

The population in 2020 was approximately 8 billion.

i Show that $\dfrac{16}{P(16 - P)} = \dfrac{1}{P} + \dfrac{1}{16 - P}$.   1 mark

ii Hence, find the solution to the logistic equation in terms of P, satisfying the initial   4 marks
conditions given above. Note that $P > 0$ and $16 - P > 0$.

iii How long (to the nearest year) will it take for the population of the world to reach   2 marks
10 billion?

c The shaded region below is bounded by the graphs of $xy = 1$, the lines $x = 2$, $y = 2$ and   4 marks
the axes. A solid is formed by rotating the shaded region about the x-axis.

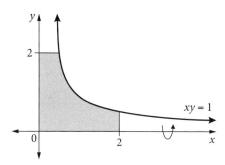

Find the exact volume of the solid.

END OF PAPER

SECTION II EXTRA WRITING SPACE

WORKED SOLUTIONS

Section I (1 mark each)

Question 1

C From the field, when $x = 0$ (y-axis),

$\dfrac{dy}{dx} = 0$, so it is options A or C, not B or D because $x = 0$ would make $\dfrac{dy}{dx}$ undefined.

When $x, y > 0$ (1st quadrant), $\dfrac{dy}{dx} < 0$, so it must be option C.

This is a common type of question.

Question 2

A $V = \pi \displaystyle\int_a^b x^2\, dy = \pi \displaystyle\int_0^1 y\, dy$

This is a straightforward question, even though some of the other options look convincing.

Question 3

A $\dfrac{dy}{dx} = 2xy$

$\displaystyle\int \dfrac{dy}{y} = \int 2x\, dx$

$\ln|y| = x^2 + c$

$|y| = e^{x^2 + c}$

This question asks for 'a' solution, so we can remove the absolute value brackets and make $c = 0$ to give equation A. This question can also be solved by differentiating every option.

Question 4

B $\dfrac{dy}{dx} = 0$ when $x = y$ (flat dashes on the line $y = x$), so options B or C.

Also, when $x < 0$, $y > 0$ (2nd quadrant),

$\dfrac{dy}{dx} > 0$ (positive gradients), so option B.

This is a common type of question.

Question 5

A Differentiating the given function:

$\dfrac{dy}{dx} = \dfrac{1}{2}(1 - x^2)^{-\frac{1}{2}}(-2x)$

$\qquad = \dfrac{-x}{\sqrt{1 - x^2}}$

$x = 0 \to y = 1 \to \dfrac{dy}{dx} = 0$ \qquad (options A or C)

$x = 1 \to y = 0 \to \dfrac{dy}{dx}$ is undefined (options A or B)

Also notice that the derivative above is in the form $-\dfrac{x}{y}$, which is option A.

There are many ways of answering this question. One other way is to notice that we can substitute y into our derivative to obtain our result.

Section II ($\checkmark$ = 1 mark)

Question 6 (15 marks)

a i $\dfrac{dy}{dx} = 0$

$y + 1 = 0$

$\qquad y = -1$ $\checkmark$

> This solution can also be identified by inspection.

ii

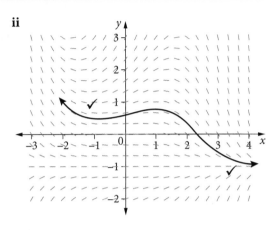

As $x \to \infty$, $y \to -1$. $\checkmark$

> Be very careful and clear when drawing solutions, ensuring to follow the field and aim for a single clean continuous line.

b

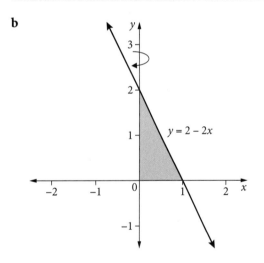

> Draw a sketch to ensure you have the correct bounds.

$V = \pi \displaystyle\int_0^2 x^2 \, dy$

$y = 2 - 2x$

$x = \dfrac{1}{2}(2 - y)$

$x^2 = \dfrac{1}{4}(2 - y)^2$

$\qquad = \dfrac{1}{4}(4 - 4y + y^2)$ $\checkmark$

$\therefore V = \dfrac{\pi}{4} \displaystyle\int_0^2 (4 - 4y + y^2) \, dy$

$\qquad = \dfrac{\pi}{4} \left[4y - 2y^2 + \dfrac{1}{3}y^3 \right]_0^2$ $\checkmark$

$\qquad = \dfrac{\pi}{4} \left[4(2) - 2(2)^2 + \dfrac{1}{3}(2)^3 - 0 \right]$

$\qquad = \dfrac{2\pi}{3}$ units^3 $\checkmark$

c $\dfrac{dy}{dx} = \sec 2y \to \dfrac{dx}{dy} = \cos 2y$ $\checkmark$

So $x = \dfrac{1}{2}\sin 2y + c.$ $\checkmark$

> This is a straightforward question. Don't try to integrate $\sec 2y$.

d $\dfrac{dP}{dt} = -\dfrac{P}{t} \to \displaystyle\int \dfrac{dP}{P} = -\int \dfrac{dt}{t}$ $\checkmark$

$\therefore \ln P = -\ln t + c,\ P > 0,\ t > 0$ $\checkmark$

Hence, $P = e^{-\ln t + c} = \dfrac{A}{t}$, where $A = e^c$. $\checkmark$

> This is a straightforward question.

e $\dfrac{dy}{dx} = 2x(1 + y^2) \to \displaystyle\int \dfrac{dy}{1 + y^2} = \int 2x \, dx$

$\therefore \tan^{-1} y = x^2 + c$ $\checkmark$

When $x = 0,\ y = 1$:

$\tan^{-1} 1 = 0^2 + c$

$\qquad c = \dfrac{\pi}{4}$ $\checkmark$

$\therefore \tan^{-1} y = x^2 + \dfrac{\pi}{4}$

$\qquad y = \tan\left(x^2 + \dfrac{\pi}{4} \right)$ $\checkmark$

> Recognise integrals that become inverse tan functions.

Question 7 (15 marks)

a Find points of intersection for limits
of integration:

$$\sqrt{x} = \frac{1}{2}x$$

$$x = \frac{1}{4}x^2$$

$$4x = x^2$$

$$0 = x^2 - 4x$$

$$0 = x(x - 4)$$

$$x = 0, 4 \checkmark$$

$$V = \pi \int_0^4 \left(y_1^2 - y_2^2\right) dx$$

$$= \pi \int_0^4 \left[\left(\sqrt{x}\right)^2 - \left(\frac{1}{2}x\right)^2\right] dx$$

$$= \pi \int_0^4 \left(x - \frac{1}{4}x^2\right) dx \checkmark$$

$$= \pi \left[\frac{1}{2}x^2 - \frac{1}{12}x^3\right]_0^4 \checkmark$$

$$= \pi \left[\frac{1}{2}(4)^2 - \frac{1}{12}(4)^3 - 0\right]$$

$$= \frac{8\pi}{3} \text{ units}^3 \checkmark$$

Areas or volumes between curves is a common
type of question. Do NOT use $(y_1 - y_2)^2$, a common
mistake.

b i $\text{RHS} = \dfrac{1}{P} + \dfrac{1}{16 - P}$

$$= \frac{16 - P}{P(16 - P)} + \frac{P}{(16 - P)P}$$

$$= \frac{16}{P(16 - P)}$$

$$= \text{LHS} \checkmark$$

It is easier to go from RHS to LHS. Going from
LHS to RHS is just as viable, but may take a bit
more work in establishing partial fractions.

ii $\dfrac{dP}{dt} = \dfrac{1}{800}P(16 - P)$

$$\frac{dt}{dP} = \frac{800}{P(16 - P)}$$

$$= 50\left(\frac{1}{P} + \frac{1}{16 - P}\right) \quad \text{(from part i)} \checkmark$$

Hence, $t = 50(\ln P - \ln[16 - P]) + c$

$$= 50\ln\left(\frac{P}{16 - P}\right) + c,$$

$$\text{as } P > 0, 16 - P > 0.$$

When $t = 0$, $P = 8$:

$$0 = 50\ln 1 + c$$

$$= 0 + c$$

Hence, $t = 50\ln\left(\dfrac{P}{16 - P}\right)$. $\checkmark$

Thus, $\dfrac{P}{16 - P} = e^{\frac{t}{50}}$

$$P = e^{\frac{t}{50}}(16 - P)$$

$$P(1 + e^{\frac{t}{50}}) = 16e^{\frac{t}{50}} \checkmark$$

$$\therefore P = \frac{16e^{\frac{t}{50}}}{1 + e^{\frac{t}{50}}}$$

$$= \frac{16}{1 + e^{-\frac{t}{50}}} \checkmark$$

The logistic equation is a common type of
question. Try to leave only one exponential in
the answer.

iii $\qquad 10 = \dfrac{16}{1 + e^{-\frac{t}{50}}}$

$$1 + e^{-\frac{t}{50}} = \frac{16}{10}$$

$$e^{-\frac{t}{50}} = \frac{6}{10} \checkmark$$

$$\therefore -\frac{t}{50} = \ln\left(\frac{6}{10}\right)$$

$$t = -50\ln\left(\frac{6}{10}\right)$$

$$\approx 25.54$$

It will take about 26 years after 2020. $\checkmark$

Make sure you answer the question in words,
giving units.

c The region is made up of 2 parts: a tall rectangle on the left and the area under the curve on the right.

When $y = 2$, $x = \dfrac{1}{2}$.

So $y = 2$ for $0 < x < \dfrac{1}{2}$

$xy = 1$ for $\dfrac{1}{2} < x < 2$. ✔

Hence, $V = \pi \displaystyle\int_0^{\frac{1}{2}} 2^2 \, dx + \pi \int_{\frac{1}{2}}^2 \dfrac{1}{x^2} \, dx$ ✔

$= \pi \left[4x \right]_0^{\frac{1}{2}} + \pi \left[-\dfrac{1}{x} \right]_{\frac{1}{2}}^2$ ✔

$= \pi \left[4\left(\dfrac{1}{2} \right) - 0 \right] + \pi \left[-\dfrac{1}{2} + \dfrac{1}{\frac{1}{2}} \right]$

$= 2\pi + \dfrac{3\pi}{2}$

$= \dfrac{7\pi}{2}$ units^3. ✔

This is a complex question. Make sure to separate into 2 volumes. The first volume can also be found by the cylinder formula where $r = 2$ and $h = \dfrac{1}{2}$.

HSC exam topic grid (2011–2020)

This table shows the coverage of this topic in past HSC exams by question number. The past exams can be downloaded from the NESA website (www.educationstandards.nsw.edu.au) by selecting 'Year 11 – Year 12', 'HSC exam papers'. NESA marking feedback and guidelines can also be found there.

The new Mathematics Extension 1 course was first examined in 2020. For exams before 2020, select 'Year 11 – Year 12', 'Resources archive', 'HSC exam papers archive'.

Differential equations were introduced to the Mathematics Extension 1 course in 2020, although similar questions have been asked in previous years.

	Solving differential equations	Direction fields	Applications of differential equations	Exponential growth and decay
2011	4(c) Maths Extension 2 exam			5(b)
2012				
2013				12(c)
2014				12(f)
2015	2			
2016				12(b)
2017			14(c)	
2018				5
2019	12(d)			12(d)
2020 new course	11(e), 12(e)	7		

Note: For volumes of integration questions, see the topic grid on page 49 of Chapter 4.

CHAPTER 6
TOPIC EXAM

6

The binomial distribution

ME-S1 The binomial distribution

S1.1 Bernoulli and binomial distributions

S1.2 Normal approximation for the sample proportion

• A reference sheet is provided on page 155 at the back of this book. • For questions in Section II, show relevant mathematical reasoning and/or calculations. **Section I – 5 questions, 5 marks** • Attempt Questions 1–5 • Allow about 8 minutes for this section **Section II – 2 questions, 30 marks** • Attempt Questions 6–7 • Allow about 52 minutes for this section	**Reading time: 5 minutes** **Working time: 1 hour** **Total marks: 35**

Section I

Question 1

Angela has a 60% chance of hitting a target.

What is the approximate probability that she hits it exactly 2 times in 5 attempts?

A 23%   B 35%

C 36%   D 58%

Question 2

Let X be the number of 1s obtained when a six-sided die is rolled 10 times.

What is the standard deviation of the distribution of X, correct to one decimal place?

A 1.2   B 1.3

C 1.4   D 1.7

Question 3

A fair coin is flipped 400 times.

What is the approximate probability that it turns up heads at least 220 times?

A 0.15%   B 0.3%

C 2.5%   D 5%

Question 4

If $X \sim \text{Bin}(n, p)$ where $E(X) = 10$ and $\text{Var}(X) = 2$, what is the value of p?

A 0.2   B 0.4

C 0.6   D 0.8

Question 5

Let X be the number of successes in n Bernoulli trials, where the probability of a success is $\frac{1}{5}$.

What is n if the standard deviation of the distribution of X is 4?

A 20   B 25

C 80   D 100

Section II

> - Attempt Questions 6–7   **30 marks**
> - Allow about 52 minutes for this section
> - Answer the questions in the spaces provided. These spaces provide
>   guidance for the expected length of response.
> - Your responses should include relevant mathematical reasoning
>   and/or calculations.

Question 6 (15 marks)

a   A six-sided die is rolled 12 times.

Let X be the number of 6s rolled.

   i   Show that $E(X)$, the expected number of 6s, is 2.   1 mark

   ii   Find, correct to four decimal places, the probability of rolling exactly the expected   2 marks
number of 6s.

   iii   Find, correct to four decimal places, the probability of rolling less than the expected   2 marks
number of 6s.

b Emily randomly guesses the answers to 10 multiple-choice questions. Each question has 4 possible answers.

   i Let X represent the number of questions Emily guesses correctly. The distribution of X is $X \sim \text{Bin}(n, p)$.

     Find the values of n and p.

                                      1 mark

   ii Calculate the values of $E(X)$ and $\text{Var}(X)$.   2 marks

   iii If the number of questions Emily guesses correctly is 1 standard deviation above the mean (to the nearest whole number), will she answer more than half of the questions correctly?   1 mark

c A fair coin is flipped 7 times.

   Let X represent the number of tails obtained.

   i Explain in words why $P(X \text{ is odd}) = \dfrac{1}{2}$.   1 mark

   ii Hence, find $P(X \text{ is prime})$.   2 marks

d X is a binomial variable with $X \sim \text{Bin}\left(n, \frac{1}{2}\right)$.   3 marks

Prove that $E(X^2) = \frac{1}{4}n(n + 1)$.

Question 7 (15 marks)

a In a recent election, 52% of people voted for the Woo Party. A survey of 1400 people was taken just before the election.

i Find the mean and standard deviation of the sample proportion of the survey, correct to three significant figures.   2 marks

ii The survey showed that only 48% of people intended to vote for the Woo Party.   2 marks

Explain why the probability of this result is approximately 0.15%.

b Elissa wants to be sure her six-sided die is fair. To test this, she will roll it a number of   3 marks
times and count the number of 6s that turn up.

Assuming that the die is fair, find the size of the samples Elissa requires so that the
standard deviation of the sample proportion of 6s is less than 1%.

c It is known that 2% of a population have green eyes. A sample of 1000 subjects is randomly   4 marks
chosen from the population. The extract below is from a table that gives $P(Z < z)$, where Z
has a standard normal distribution.

Use the table to find the probability that no more than 3% of the sample have green eyes.
Write the answer as a percentage correct to one decimal place.

z	0.00	0.01	0.02	0.03	0.04	0.05	0.06	0.07	0.08	0.09
2.0	0.9772	0.9778	0.9783	0.9788	0.9793	0.9798	0.9803	0.9808	0.9812	0.9817
2.1	0.9821	0.9826	0.9830	0.9834	0.9838	0.9842	0.9846	0.9850	0.9854	0.9857
2.2	0.9861	0.9864	0.9868	0.9871	0.9875	0.9878	0.9881	0.9884	0.9887	0.9890
2.3	0.9893	0.9896	0.9898	0.9901	0.9904	0.9906	0.9909	0.9911	0.9913	0.9916
2.4	0.9918	0.9920	0.9922	0.9925	0.9927	0.9929	0.9931	0.9932	0.9934	0.9936

d In a computer game, the probability of obtaining a pearl when bartering with a character
   is 5%. A player barters with this character 262 times.

    i Find the mean and standard deviation of the sample proportion of pearls   2 marks
   (standard deviation correct to four significant figures).

    ii If $P(X > 8) < 10^{-15}$, where X has a standard normal distribution, explain why there is   2 marks
   a less than 10^{-15} chance that a player could obtain 42 pearls.

END OF PAPER

TOPIC EXAM

WORKED SOLUTIONS

Section I (1 mark each)

Question 1

A $n = 5, p = 0.6$

$$P(X = 2) = \binom{5}{2}(0.6)^2(0.4)^3 \approx 0.2304$$

This is a common type of binomial probability question.

Question 2

A $n = 10, p = \dfrac{1}{6}$

$$\text{Var}(X) = 10 \times \frac{1}{6} \times \frac{5}{6}$$

$$= \frac{25}{18}$$

So $\sigma = \sqrt{\dfrac{25}{18}} \approx 1.2$.

This is a straightforward binomial distribution question using the Var(X) formula from the HSC exam reference sheet.

Question 3

C $n = 400, p = \dfrac{1}{2}$

$$\text{Var}(X) = 400 \times \frac{1}{2} \times \frac{1}{2} = 100$$

$$\sigma = \sqrt{100}$$
$$= 10$$

$$\mu = E(X) = 400 \times \frac{1}{2} = 200$$

220 is 2 standard deviations above the mean.

$$220 = \mu + 2\sigma$$

So $P(X \geq 220) \approx \dfrac{1}{2}(100\% - 95\%)$

$$= 2.5\%.$$

You should be able to recognise the normal approximations to the binomial distribution.

Question 4

D $np = 10$ and $np(1 - p) = 2$

Divide the 2nd equation by the 1st equation (or substitute the 1st into the 2nd).

$$\therefore 1 - p = \frac{1}{5}$$

$$p = \frac{4}{5}$$

This is a straightforward question.

Question 5

D $p = \dfrac{1}{5}, \sigma = 4$

$$\text{Var}(X) = 4^2 = 16$$

But $\text{Var}(X) = np(1 - p)$

$$16 = n \times \frac{1}{5} \times \frac{4}{5}$$

$$16 = \frac{4n}{25}$$

So $n = 100$.

These types of questions use probability formulas to set up and solve algebraic equations.

Section II (✓ = 1 mark)

Question 6 (15 marks)

a i $X \sim \text{Bin}\left(12, \dfrac{1}{6}\right)$

So $E(X) = np = 12 \times \dfrac{1}{6} = 2$. ✓

This is a straightforward question.
Use the HSC exam reference sheet.

ii $P(X = 2) = \dbinom{12}{2}\left(\dfrac{1}{6}\right)^2\left(\dfrac{5}{6}\right)^{10}$ ✓

$$\approx 0.2961 ✓$$

This is a common type of binomial probability question.

iii $P(X < 2) = P(X = 0) + P(X = 1)$

$$= \binom{12}{0}\left(\frac{1}{6}\right)^0\left(\frac{5}{6}\right)^{12} + \binom{12}{1}\left(\frac{1}{6}\right)^1\left(\frac{5}{6}\right)^{11} ✓$$

$$\approx 0.3813 ✓$$

This is a common type of question.

b i $n = 10$ and $p = \dfrac{1}{4}$ ✓

This is another straightforward question.

WORKED SOLUTIONS

ii $E(X) = np = 10 \times \dfrac{1}{4} = 2.5$ ✓

$\text{Var}(X) = np(1 - p) = 10 \times \dfrac{1}{4} \times \dfrac{3}{4} = 1.875$ ✓

Make sure you use the HSC exam reference sheet.

iii $\sigma = \sqrt{1.875} \approx 1.369$

$\mu + \sigma \approx 2.5 + 1.369 = 3.869 \approx 4 < 5$

No, Emily will not answer more than half of the questions correctly. ✓

Make sure you provide reasoning, and not just a bare answer.

c i For each trial, the probability of flipping a tail is $\dfrac{1}{2}$, the same as the probability of flipping a head.

Odd is 1, 3, 5 or 7 tails. Even is 0, 2, 4 or 6 tails; that is, 7, 5, 3 or 1 heads.

Since the situation or distribution is symmetrical, and probability of flipping a tail is equal to its complement, the probability of flipping an odd number of tails must be $\dfrac{1}{2}$. ✓

This is only 1 mark, so that should indicate that you don't have to calculate the probability.

ii $P(X \text{ is prime})$
$= P(2) + P(3) + P(5) + P(7)$
$= P(\text{odd}) - P(1) + P(2)$
$= \dfrac{1}{2} - \dbinom{7}{1}\left(\dfrac{1}{2}\right)^1\left(\dfrac{1}{2}\right)^6 + \dbinom{7}{2}\left(\dfrac{1}{2}\right)^2\left(\dfrac{1}{2}\right)^5$ ✓
$= \dfrac{39}{64}$ ✓

You should use part **i** to reduce the amount of working needed. Also, don't forget, 1 is NOT a prime number.

d $\text{Var}(X) = E(X^2) - \mu^2 \qquad$ (from the
$= E(X^2) - (np)^2 \qquad$ reference sheet)
$= E(X^2) - \left(\dfrac{1}{2}n\right)^2$
$= E(X^2) - \dfrac{1}{4}n^2$ ✓

But $\text{Var}(X) = np(1 - p)$
$= n \times \dfrac{1}{2} \times \dfrac{1}{2}$
$= \dfrac{1}{4}n.$ ✓

$\therefore \dfrac{1}{4}n = E(X^2) - \dfrac{1}{4}n^2$

$E(X^2) = \dfrac{1}{4}n + \dfrac{1}{4}n^2$
$= \dfrac{1}{4}n(n + 1)$ ✓

This question is an example of why the HSC exam reference sheet is so important. At first glance, this question might seem impossible, but the formula from the reference sheet makes it easy.

Question 7 (15 marks)

a i $\mu = p = 52\% = 0.52$ ✓

$\sigma = \sqrt{\dfrac{p(1 - p)}{n}}$
$= \sqrt{\dfrac{0.52 \times 0.48}{1400}}$
$\approx 0.013\,35$
≈ 0.013 ✓

This is a straightforward question.

ii 0.48 is 3 standard deviations below the mean.

$0.48 \approx 0.52 - 3 \times 0.013$
$= \mu - 3\sigma$ ✓

$\therefore P(\hat{p} \le 0.48) \approx \dfrac{1}{2}(100\% - 99.7\%)$
$\qquad\qquad$ (normal distribution)
$= 0.15\%$ ✓

Since there is no other information, it is reasonable to assume that the answer will come from the normal distribution on the HSC exam reference sheet. It is important to be able to recognise these percentages.

b $\sigma = \sqrt{\dfrac{p(1 - p)}{n}} = \sqrt{\dfrac{\frac{1}{6} \times \frac{5}{6}}{n}} \le 0.01$ ✓

Hence, $\dfrac{5}{36n} \le 0.01^2 = 0.0001.$ ✓

$\therefore \dfrac{36n}{5} \ge 10\,000$ ✓

$n \ge 1389$
$= 1389$

Sample proportion concepts and problems are complex and require deep study. The formula for the standard deviation of the sample proportion is very important to remember.

c $\mu = p = 0.02$, $n = 1000$ ✓

$$\sigma = \sqrt{\frac{0.02 \times 0.98}{1000}}$$

$$\approx 0.004\,427 \checkmark$$

$$z = \frac{x - \mu}{\sigma}$$

$$\approx \frac{0.03 - 0.02}{0.004\,427}$$

$$\approx 2.259 \checkmark$$

From the table, $P(Z < 2.26) = 0.9881$.

Hence, the probability that no more than 3% of the population has green eyes is approximately 98.8%. ✓

This is a straightforward question but you need to practise the process.

d i $\mu = p = 5\% = 0.05$ ✓

$$\sigma = \sqrt{\frac{p(1-p)}{n}}$$

$$= \sqrt{\frac{0.05 \times 0.95}{262}}$$

$$\approx 0.013\,46 \checkmark$$

This is a formula question.

ii $x = \dfrac{42}{262}$

$$\approx 0.1603$$

$$z = \frac{x - \mu}{\sigma}$$

$$\approx \frac{0.1603 - 0.05}{0.013\,46}$$

$$\approx 8.2 \checkmark$$

Hence, the probability that the player could obtain 42 pearls is less than $P(Z > 8)$, assuming a normal distribution, which is less than 10^{-15}. ✓

This is a straightforward question but requires a good understanding of what is happening in the question. Fun fact: this is based on an actual computer game!

HSC exam topic grid (2011–2020)

This table shows the coverage of this topic in past HSC exams by question number. The past exams can be downloaded from the NESA website (www.educationstandards.nsw.edu.au) by selecting 'Year 11 – Year 12', 'HSC exam papers'. NESA marking feedback and guidelines can also be found there.

The new Mathematics Extension 1 course was first examined in 2020. For exams before 2020, select 'Year 11 – Year 12', 'Resources archive', 'HSC exam papers archive'.

Binomial distributions and sample proportion were introduced to the Mathematics Extension 1 course in 2020.

	Binomial probability	Binomial distributions	Sample proportion
2011			
2012			
2013	11(c)		
2014	11(b)		
2015	14(c)		
2016	11(f)		
2017	11(g)		
2018	12(d)		
2019	11(f)		
2020 new course		12(b)	

Mathematics Extension 1

PRACTICE MINI-HSC EXAM 1

General instructions	• Reading time: 5 minutes
	• Working time: 1 hour
	• A reference sheet is provided on page 155 at the back of this book
	• For questions in Section II, show relevant mathematical reasoning and/or calculations.
Total marks: 35	**Section I – 5 questions, 5 marks**
	• Attempt Questions 1–5
	• Allow about 8 minutes for this section
	Section II – 2 questions, 30 marks
	• Attempt Questions 6–7
	• Allow about 52 minutes for this section

Section I

5 marks
Attempt Questions 1–5
Allow about 8 minutes for this section

Circle the correct answer.

Question 1
Which value is a double root of $x^3 - 7x^2 + 16x - 12 = 0$?

A -2

B -1

C 1

D 2

Question 2
What is the angle between the vectors $\begin{pmatrix} -1 \\ 2 \end{pmatrix}$ and $\begin{pmatrix} -1 \\ -3 \end{pmatrix}$?

A $30°$

B $35°$

C $135°$

D $150°$

Question 3
If the area of a square is increasing at $4\,\text{cm}^2/\text{s}$, at what rate is its perimeter increasing when it is $8\,\text{cm}$?

A $1\,\text{cm/s}$

B $2\,\text{cm/s}$

C $4\,\text{cm/s}$

D $8\,\text{cm/s}$

Question 4

Which slope field is for the differential equation $\dfrac{dy}{dx} = x^2 - y$?

A

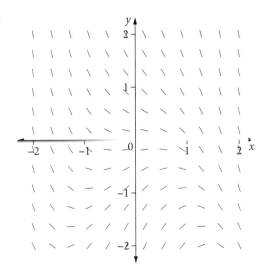

B

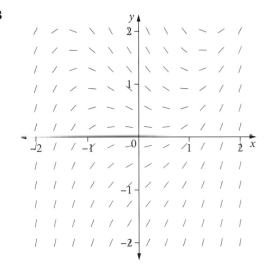

C

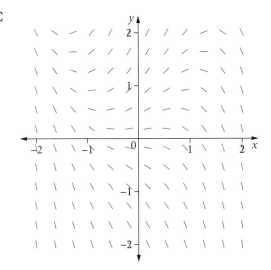

D

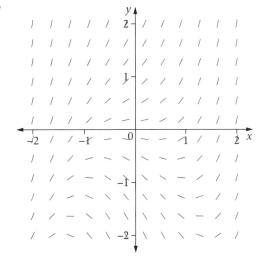

Question 5

What is the general solution to the equation $4\cos x = 1 + 4\cos^2 x$, where n is an integer?

A $x = \dfrac{\pi}{6} + 2n\pi, \dfrac{11\pi}{6} + 2n\pi$

B $x = \dfrac{\pi}{3} + 2n\pi, \dfrac{5\pi}{3} + 2n\pi$

C $x = \dfrac{2\pi}{3} + 2n\pi, \dfrac{4\pi}{3} + 2n\pi$

D $x = \dfrac{5\pi}{6} + 2n\pi, \dfrac{7\pi}{6} + 2n\pi$

Section II

30 marks
Attempt Questions 6–7
Allow about 52 minutes for this section

- Answer the questions in the spaces provided. These spaces provide guidance for the expected length of response.
- Your responses should include relevant mathematical reasoning and/or calculations.

Question 6 (15 marks)

a   Differentiate $\sin^{-1}(\tan x)$.   2 marks

b   The diagram below is a sketch of $y = f(x)$.   3 marks

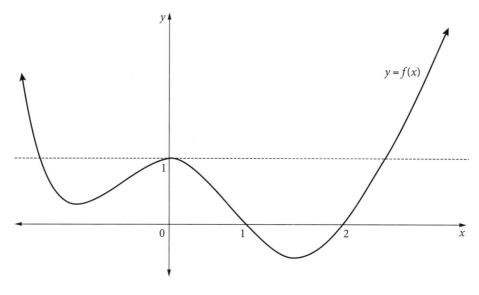

Sketch on the diagram the graph of $y = \sqrt{f(x)}$.

c   At the Year 12 camp, students can choose 3 activities from a list of 6.   2 marks

If there are 165 students in Year 12, explain why there must be at least 9 students who chose the same 3 activities.

Question 6 continues on page 77

Question 6 (continued)

d   Solve $\dfrac{dP}{dt} = -\dfrac{P}{2}$, finding P as a function of t if $P > 0$.   2 marks

e   A pair of six-sided dice are rolled and the sum of the numbers is calculated. This is done a total   3 marks
of 5 times. Let X be the number of times the sum is 8, 9 or 10.

Find $P(X \geq 3)$.

f   Use the substitution $x = u^2 - 2$ to evaluate $\displaystyle\int_{-2}^{2} x\sqrt{2 + x}\, dx$.   3 marks

End of Question 6

Question 7 (15 marks)

a   Alfie put some jelly into the fridge, which was set to a temperature of 4°C. The temperature of the jelly was initially 50°C and after 1 hour its temperature was 29°C.

The change in temperature of the jelly can be modelled by the equation $\dfrac{dJ}{dt} = -k(J - 4)$, where J is the temperature of the jelly in °C, t is the time the jelly has been in the fridge in minutes, and k is a positive constant.

   i   Show that $J = 4 + Ae^{-kt}$ satisfies the above equation, where A is a positive constant.   1 mark

   ii   How long, to the nearest hour, will it take for the temperature of the jelly to reach 5°C?   3 marks

b   The region bounded by $y = 2\cos^{-1}x$ and the axes are rotated around the y-axis to form a solid.   3 marks

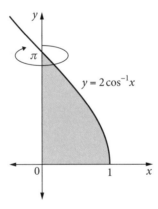

Find the volume of the solid.

Question 7 continues on page 79

Question 7 (continued)

c The diagram below shows $\triangle AOB$, where $\overrightarrow{OA} = 3\underset{\sim}{i} + 4\underset{\sim}{j}$ and $\overrightarrow{OB} = 4\underset{\sim}{i} + \underset{\sim}{j}$.

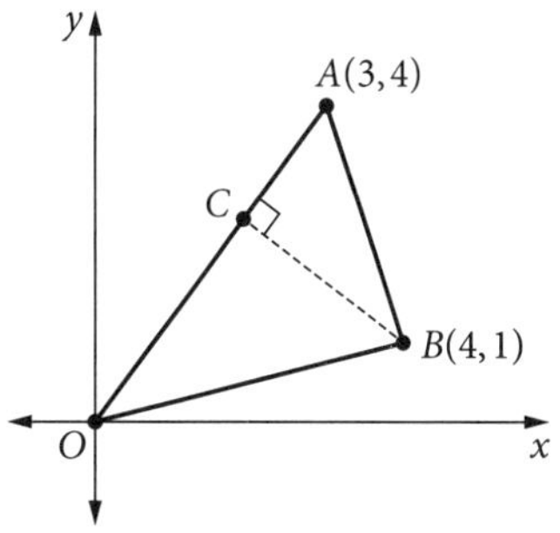

NOT TO SCALE

The point C lies on OA such that CB is perpendicular to OA.

i   By considering $\text{proj}_{\overrightarrow{OA}}\overrightarrow{OB}$, the projection of $\overrightarrow{OB}$ onto $\overrightarrow{OA}$, find the coordinates of C.   2 marks

ii Hence, calculate the exact area of $\triangle OAB$.   2 marks

Question 7 continues on page 80

Question 7 (continued)

d Prove by mathematical induction that $47^{n+1} - 188 \times 4^{n-1}$ is divisible by 2021 for all positive integers n.　　4 marks

END OF PAPER

WORKED SOLUTIONS

Section I (1 mark each)

WORKED SOLUTIONS

Question 1

D $P(x) = x^3 - 7x^2 + 16x - 12$

$P'(x) = 3x^2 - 14x + 16 = 0$

$0 = (3x - 8)(x - 2)$

So $x = \dfrac{8}{3}$ or 2.

This is a straightforward question from the Year 11 polynomials topic.

Question 2

C $\begin{pmatrix} -1 \\ 2 \end{pmatrix} \times \begin{pmatrix} -1 \\ -3 \end{pmatrix} = \left\| \begin{pmatrix} -1 \\ 2 \end{pmatrix} \right\| \left\| \begin{pmatrix} -1 \\ -3 \end{pmatrix} \right\| \cos\theta$

$1 - 6 = \sqrt{5} \times \sqrt{10} \cos\theta$

$\cos\theta = -\dfrac{1}{\sqrt{2}}$

$\theta = 135°$

Sketch a diagram to check that the angle looks correct.

Question 3

C Area: $A = s^2$, $\dfrac{dA}{dt} = 4$

Perimeter: $P = 4s$

$\dfrac{dP}{dt} = ?$ at $P = 8$

$\dfrac{dP}{dt} = \dfrac{dP}{ds} \times \dfrac{ds}{dA} \times \dfrac{dA}{dt}$

$= 4 \times \dfrac{1}{\dfrac{dA}{ds}} \times 4$

$= 16 \times \dfrac{1}{2s}$

$= \dfrac{8}{s}$

When $P = 8$, $s = \dfrac{1}{4} \times 8 = 2$.

$\dfrac{dP}{dt} = \dfrac{8}{2}$

$= 4\,\text{cm/s}$

This is a common related rates question. Having derivatives that slightly relate to each other usually indicates chain rule usage.

Question 4

B If $y < 0$, then $\dfrac{dy}{dx} > 0$, so look for positive gradients below the x-axis (3rd and 4th quadrants). Only option B has all positive dashes.

Look for features to eliminate options. Examine what happens if x or y is positive, negative or 0.

Question 5

B $4\cos x = 1 + 4\cos^2 x$

$4\cos^2 x - 4\cos x + 1 = 0$

$(2\cos x - 1)^2 = 0$

$\cos x = \dfrac{1}{2}$

$x = \dfrac{\pi}{3}, \dfrac{5\pi}{3}, \dfrac{7\pi}{3}, \dfrac{11\pi}{3}, \ldots$

Identify which answer matches the pattern of solutions.

Section II (✓ = 1 mark)

Question 6 (15 marks)

a $\dfrac{d}{dx}\sin^{-1}(\tan x) = \dfrac{\sec^2 x}{\sqrt{1 - \tan^2 x}}$ ✓✓

> Use the HSC exam reference sheet.

b

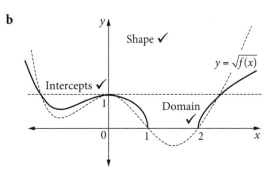

> This is a straightforward Year 11 graphing question; calculus is not required.

c The number of ways to choose 3 from 6 is

$\binom{6}{3} = 20.$ ✓

$\dfrac{165}{20} = 8.25 > 8$; hence, by the pigeonhole principle, there must be a selection of 3 activities that was chosen by at least 9 students. ✓

> This is a Year 11 combinations and pigeonhole principle question. Make sure that you practise these types of questions and give enough explanation to make it clear.

d $\dfrac{dP}{dt} = -\dfrac{P}{2}$

$\dfrac{dt}{dP} = -\dfrac{2}{P}$

$t = -2\ln P + c$ ✓ $(P > 0)$

$\therefore P = e^{\frac{c-t}{2}}$

$= Ae^{-\frac{t}{2}}$, where $A = e^{\frac{c}{2}}$ ✓

> This is a common differential equations question. Don't forget the constant term '+ c'.

e With 2 six-sided dice, there are 5 ways of making 8, 4 ways of making 9 and 3 ways of making 10.

Hence, the probability of getting a sum of 8, 9 or 10 with 2 dice is $\dfrac{12}{36} = \dfrac{1}{3}$. ✓

$P(X \geq 3)$
$= P(X = 3) + P(X = 4) + P(X = 5)$
$= \binom{5}{3}\left(\dfrac{1}{3}\right)^3\left(\dfrac{2}{3}\right)^2 + \binom{5}{4}\left(\dfrac{1}{3}\right)^4\left(\dfrac{2}{3}\right)^1 + \binom{5}{5}\left(\dfrac{1}{3}\right)^5\left(\dfrac{2}{3}\right)^0$ ✓
$= \dfrac{17}{81}$ ✓

> Binomial probability is often tested in the HSC exams. See the topic grid on page 72. Read the question carefully.

f $x = u^2 - 2$
$dx = 2u\,du$ ✓

$x = -2 \rightarrow u = 0$
$x = 2 \rightarrow u = 2$

Hence, $\displaystyle\int_{-2}^{2} x\sqrt{2 + x}\,dx = \int_{0}^{2} 2u(u^2 - 2)u\,du$

$= \displaystyle\int_{0}^{2}(2u^4 - 4u^2)\,du$ ✓

$= \left[\dfrac{2}{5}u^5 - \dfrac{4}{3}u^3\right]_0^2$

$= \dfrac{2}{5}(2)^5 - \dfrac{4}{3}(2)^3 - 0$

$= 2\dfrac{2}{15}$ ✓

> This is a common type of exam question; definite integration by substitution. Make sure that you adjust the limits of integration.

Question 7 (15 marks)

a **i** $\dfrac{dJ}{dt} = -kAe^{-kt}$

$-k(J - 4) = -k(4 + Ae^{-kt} - 4)$
$= -kAe^{-kt}$

So $\dfrac{dJ}{dt} = -k(J - 4).$ ✓

> This is a straightforward and common type of question on Newton's law of cooling.

ii When $t = 0$, $J = 50$:

$$50 = 4 + Ae^{-k(0)}$$
$$A = 46 \checkmark$$

$$J = 4 + 46e^{-kt}$$

When $t = 60$, $J = 29$:

$$29 = 4 + 46e^{-k(60)}$$
$$25 = 46e^{-60k}$$
$$e^{-60k} = \frac{25}{46}$$
$$-60k = \ln\left(\frac{25}{46}\right)$$
$$k = -\frac{1}{60}\ln\left(\frac{25}{46}\right)$$
$$\approx 0.010\,16 \checkmark$$

When $J = 5$:

$$5 = 4 + 46e^{-kt}$$
$$1 = 46e^{-kt}$$
$$e^{-kt} = \frac{1}{46}$$
$$-kt = \ln\left(\frac{1}{46}\right)$$
$$t = -\frac{1}{k}\ln\left(\frac{1}{46}\right)$$
$$\approx 376.7 \text{ minutes}$$
$$= 6\,\text{h}\,16.7\,\text{min}$$
$$\approx 6 \text{ hours} \checkmark$$

It will take approximately 6 hours for the jelly to reach 5°C.

b $y = 2\cos^{-1} x$

$$x = \cos\left(\frac{y}{2}\right) \checkmark$$

$$V = \pi \int_0^{\pi} \cos^2\left(\frac{y}{2}\right) dy$$

$$= \pi \int_0^{\pi} \frac{1}{2}(\cos y + 1)\, dy \quad \checkmark \quad \text{(from the reference sheet)}$$

$$= \frac{\pi}{2}\left[\sin y + y\right]_0^{\pi}$$

$$= \frac{\pi}{2}\left[\sin \pi + \pi - 0\right]$$

$$= \frac{\pi^2}{2} \text{ units}^3 \checkmark$$

This is a typical volumes question that requires the use of trigonometric identities.

c i $\overrightarrow{OC} = \text{proj}_{\overrightarrow{OA}}\overrightarrow{OB}$

$$= \frac{(3\underset{\sim}{i} + 4\underset{\sim}{j}) \cdot (4\underset{\sim}{i} + \underset{\sim}{j})}{(3\underset{\sim}{i} + 4\underset{\sim}{j}) \cdot (3\underset{\sim}{i} + 4\underset{\sim}{j})}(3\underset{\sim}{i} + 4\underset{\sim}{j}) \quad \checkmark$$

$$= \frac{12 + 4}{9 + 16}(3\underset{\sim}{i} + 4\underset{\sim}{j})$$

$$= \frac{16}{25}(3\underset{\sim}{i} + 4\underset{\sim}{j})$$

$$= 1.92\underset{\sim}{i} + 2.56\underset{\sim}{j}$$

So the coordinates of C are $(1.92, 2.56)$. $\checkmark$

Make sure you know what vector projection is and its formula.

ii $|BC| = \sqrt{(1.92 - 4)^2 + (2.56 - 1)^2}$

$$= \sqrt{2.08^2 + 1.56^2}$$

$$= \sqrt{6.76}$$

$$= 2.6$$

$$|OA| = \sqrt{3^2 + 4^2} = 5 \checkmark$$

Area of $\triangle OAB = \frac{1}{2} \times 2.6 \times 5$

$$= 6.5 \text{ units}^2 \checkmark$$

This is a straightforward question that is linked to part **i**.

d Let $P(n)$ be the statement $47^{n+1} - 188 \times 4^{n-1}$ is divisible by 2021.

$P(1)$ is $47^2 - 188 \times 4^0 = 2021$, which is divisible by 2021.

$P(1)$ is true. ✓

Assume $P(k)$ is true.

Thus, $47^{k+1} - 188 \times 4^{k-1} = 2021p$ for some integer p.

$\therefore 47^{k+1} = 2021p + 188 \times 4^{k-1}$ [*] ✓

RTP $P(k+1)$: $47^{k+2} - 188 \times 4^k$ is divisible by 2021.

$47^{k+2} - 188 \times 4^k$

$= 47(47^{k+1}) - 188 \times 4^k$

$= 47(2021p + 188 \times 4^{k-1}) - 188 \times 4^k$ by [*] ✓

$= 2021(47p) + 8836(4^{k-1}) - 188(4^{k-1} \times 4)$

$= 2021(47p) + 8836(4^{k-1}) - 752(4^{k-1})$

$= 2021(47p) + 8084(4^{k-1})$

$= 2021[47p + 4(4^{k-1})]$, which is divisible by 2021.

$\therefore P(k+1)$ is true.

So by mathematical induction, $P(n)$ is true for all positive integers n. ✓

> There are other ways to do this challenging proof, but they depend on being very careful with the inductive step and manipulating powers properly.

Mathematics Extension 1

PRACTICE MINI-HSC EXAM 2

General instructions	• Reading time: 5 minutes
	• Working time: 1 hour
	• A reference sheet is provided on page 155 at the back of this book
	• For questions in Section II, show relevant mathematical reasoning and/or calculations.

Total marks: 35	**Section I – 5 questions, 5 marks**
	• Attempt Questions 1–5
	• Allow about 8 minutes for this section
	Section II – 2 questions, 30 marks
	• Attempt Questions 6–7
	• Allow about 52 minutes for this section

Section I

5 marks
Attempt Questions 1–5
Allow about 8 minutes for this section

Circle the correct answer.

Question 1

How many ways can 4 boys and 4 girls sit around a circle if boys and girls alternate?

A   144

B   576

C   5040

D   40 320

Question 2

Find the derivative of $\cos^{-1}\left(\dfrac{x}{3}\right)$.

A $-\dfrac{3}{\sqrt{9-x^2}}$

B $-\dfrac{1}{\sqrt{9-x^2}}$

C $\dfrac{1}{\sqrt{9-x^2}}$

D $\dfrac{3}{\sqrt{9-x^2}}$

Question 3

If $\cos\phi = \dfrac{8}{17}$, what is the value of $\cos 2\phi$?

A $-\dfrac{240}{289}$

B $-\dfrac{161}{289}$

C $\dfrac{161}{289}$

D $\dfrac{240}{289}$

Question 4

Which differential equation has the direction field shown?

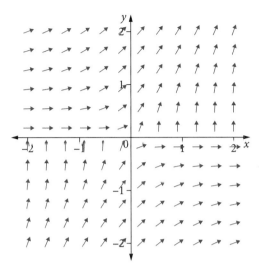

A $\dfrac{dy}{dx} = e^{\frac{x}{y}}$

B $\dfrac{dy}{dx} = e^{xy}$

C $\dfrac{dy}{dx} = e^{x-y}$

D $\dfrac{dy}{dx} = e^{x+y}$

Question 5

If $\underset{\sim}{w} = \text{proj}_{\underset{\sim}{u}}\underset{\sim}{v}$ (the projection of $\underset{\sim}{v}$ on $\underset{\sim}{u}$), then which equation is true?

A $\left|\underset{\sim}{u}\right|^2 = \left|\underset{\sim}{v}\right|^2 + \left|\underset{\sim}{u} - \underset{\sim}{v}\right|^2$

B $\left|\underset{\sim}{v}\right|^2 = \left|\underset{\sim}{w}\right|^2 + \left|\underset{\sim}{v} - \underset{\sim}{w}\right|^2$

C $\left|\underset{\sim}{w}\right|^2 = \left|\underset{\sim}{v}\right|^2 + \left|\underset{\sim}{v} - \underset{\sim}{w}\right|^2$

D $\left|\underset{\sim}{w}\right|^2 = \left|\underset{\sim}{u}\right|^2 + \left|\underset{\sim}{u} - \underset{\sim}{w}\right|^2$

Section II

30 marks
Attempt Questions 6–7
Allow about 52 minutes for this section

- Answer the questions in the spaces provided. These spaces provide guidance for the expected length of response.
- Your responses should include relevant mathematical reasoning and/or calculations.

Question 6 (15 marks)

a   What is the value of the constant term in $\left(x^2 + \dfrac{2}{x}\right)^6$?   2 marks

b   Express $\sqrt{3}\sin x + \cos x$ in the form $R\sin(x + \alpha)$, where $0 < \alpha < \dfrac{\pi}{2}$.   2 marks

Question 6 continues on page 89

Question 6 (continued)

c   For what values of x is $\dfrac{x}{x+3} > 2$?   3 marks

d   Find $\int \cos^2 2x \, dx$.   2 marks

e   Solve the differential equation $\dfrac{dy}{dx} = xy - y$, given that the graph of $y = f(x)$ goes through   3 marks
the point $(1, 1)$.

Question 6 continues on page 90

Question 6 (continued)

f   Use the substitution $u = x^3$ to evaluate $\int_0^1 \frac{x^2}{1 + x^6}\, dx$.   3 marks

Question 7 (15 marks)

a   A pizza restaurant claims that more than 90% of its orders arrive within 30 minutes.

    i   A consumer group tests this on samples of 100 orders over a year.   2 marks

       Show that the mean and standard deviation of the sample proportion of the orders arriving within 30 minutes are 0.9 and 0.03, respectively.

    ii   The consumer group found that for one sample, only 81 orders arrived within 30 minutes.   2 marks

       By using a normal distribution, find the probability of this happening.

Question 7 continues on page 91

Question 7 (continued)

b Prove by mathematical induction that, for all integers $n \geq 1$,   3 marks

$$(1 \times 2 \times 3) + (2 \times 3 \times 4) + \ldots + n(n + 1)(n + 2) = \frac{1}{4}n(n + 1)(n + 2)(n + 3).$$

c The polynomial $P(x) = x^4 - x^3 + kx^2 + 25x + 300$ has 4 real roots, 2 of which have the same   4 marks
magnitude but opposite sign.

Find k.

Question 7 continues on page 92

Question 7 (continued)

d   A stunt car on a film set drives horizontally off a vertical cliff at u m/s, and lands on the ground   4 marks
at an angle of 45° to the horizontal, 90 m away from the base of the cliff.

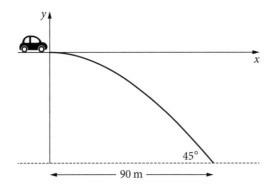

The path of the car is shown on the coordinate plane, where x is the horizontal distance (metres)
and y is the vertical distance (metres) from the point where the car leaves the cliff. The position

of the car after t seconds is given by $\underset{\sim}{r} = \begin{pmatrix} ut \\ -5t^2 \end{pmatrix}$. Do NOT prove this.

Find u.

END OF PAPER

WORKED SOLUTIONS

Section I (1 mark each)

Question 1

A

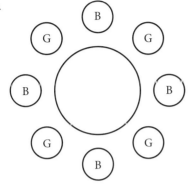

The first person can sit anywhere (boy or girl), so we just need to arrange the other 7 people around the circle, relative to the first person.

$$3! \times 4! = 144$$

This is a Year 11 permutations question.

Question 2

B $\dfrac{d}{dx} \cos^{-1}\left(\dfrac{x}{3}\right) = -\dfrac{1}{\sqrt{9 - x^2}}$

This is a common type of question.

Question 3

B $\cos 2\phi = 2\cos^2 \phi - 1$

$$= 2\left(\frac{8}{17}\right)^2 - 1$$

$$= -\frac{161}{289}$$

This question uses trigonometric identities from the HSC exam reference sheet.

Question 4

A Steep gradients in 1st and 3rd quadrants, when x and y have the same sign. Low or 0 gradients in 2nd and 4th quadrants, when x and y have different signs. So A or B as e^+ is high and e^- is low. At $(2, 2)$, A gives $e^1 = 2.7$ while B gives $e^4 = 54.6$, and dash looks more like a gradient of 2.7.

This is a hard question. Be very careful looking for features in the field. Looking for lines of constant gradient is a good strategy.

Question 5

B $\underset{\sim}{w}$ creates a right-angled triangle with $\underset{\sim}{v}$ and $\underset{\sim}{v} - \underset{\sim}{w}$ as the shorter sides, and with $\underset{\sim}{v}$ as the hypotenuse. It is important to understand the geometric significance behind projections.

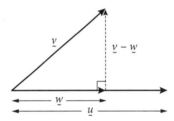

Sketching a diagram can help verify this application of Pythagoras' theorem.

Section II ($\checkmark$ = 1 mark)

Question 6 (15 marks)

a Terms of $\left(x^2 + \dfrac{2}{x}\right)^6$:

$$\binom{6}{k}(x^2)^k\left(\frac{2}{x}\right)^{6-k} = \binom{6}{k}x^{2k}2^{6-k}x^{k-6}$$

$$= \binom{6}{k}2^{6-k}x^{3k-6} \checkmark$$

We want x^0, so $3k - 6 = 0$.

Hence, $k = 2$ and the constant term:

$$\binom{6}{2}2^4 = 240. \checkmark$$

b Let $\sqrt{3}\sin x + \cos x$
$$= R\sin(x + \alpha)$$
$$= R(\sin x \cos\alpha + \cos x \sin\alpha)$$
$$= R\sin x \cos\alpha + R\cos x \sin\alpha$$

$$\therefore R\cos\alpha = \sqrt{3} \quad [1]$$
$$\therefore R\sin\alpha = 1 \quad\quad [2]$$

$$[2] \div [1]: \quad \tan\alpha = \frac{1}{\sqrt{3}}$$

$$\alpha = \frac{\pi}{6} \checkmark$$

$$[1]^2 + [2]^2: \quad R^2\cos^2\alpha + R^2\sin^2\alpha = \left(\sqrt{3}\right)^2 + 1^2$$
$$R^2(\cos^2\alpha + \sin^2\alpha) = 4$$
$$R = 2$$

$$\therefore \sqrt{3}\sin x + \cos x = 2\sin\left(x + \frac{\pi}{6}\right) \checkmark$$

This is a very common type of exam question.

c $$\frac{x}{x + 3} > 2$$
$$x(x + 3) > 2(x + 3)^2$$
$$x^2 + 3x > 2x^2 + 12x + 18$$
$$x^2 + 9x + 18 < 0 \checkmark$$
$$(x + 3)(x + 6) < 0 \checkmark$$

So $-6 < x < -3$. $\checkmark$

This is a common type of Year 11 question. You should draw a simple sketch to ensure you get the right inequalities from the factorised quadratic.

d $\displaystyle\int \cos^2 2x\, dx$

$$= \int \frac{1}{2}(1 + \cos 4x)\, dx \checkmark \quad \text{(from the reference sheet)}$$

$$= \frac{1}{2}\left(x + \frac{1}{4}\sin 4x\right) + c$$

$$= \frac{x}{2} + \frac{1}{8}\sin 4x + c \checkmark$$

Integrating the square of a sine or cosine function is a common type of Maths Extension 1 question. Use the HSC exam reference sheet and don't forget the constant term '+ c'.

e $$\frac{dy}{dx} = xy - y$$
$$= y(x - 1)$$
$$\int\frac{dy}{y} = \int(x - 1)\, dx$$

Hence, $\ln|y| = \dfrac{1}{2}x^2 - x + c$. $\checkmark$

$x = 1, y = 1$:

$$\ln 1 = \frac{1}{2}(1)^2 - 1 + c$$

$$0 = -\frac{1}{2} + c$$

$$c = \frac{1}{2} \checkmark$$

Hence, $\ln|y| = \dfrac{1}{2}x^2 - x + \dfrac{1}{2}$

$$= \frac{1}{2}(x^2 - 2x + 1)$$

$$= \frac{1}{2}(x - 1)^2.$$

So $y = e^{\frac{1}{2}(x-1)^2}$. (positive y because of $(1, 1)$) $\checkmark$

This is a straightforward question. Make sure you find c and identify whether the absolute value symbols can be removed

f $u = x^3$
$$du = 3x^2\, dx$$
$$x^2\, dx = \frac{1}{3}du \checkmark$$

When $x = 0$, $u = 0$.

When $x = 1$, $u = 1$.

Hence, $\displaystyle\int_0^1 \frac{x^2}{1 + x^6}\, dx = \frac{1}{3}\int_0^1 \frac{1}{1 + u^2}\, du \checkmark$

$$= \frac{1}{3}\left[\tan^{-1}u\right]_0^1$$

$$= \frac{1}{3}[\tan^{-1}1 - 0]$$

$$= \frac{\pi}{12}. \checkmark$$

This is a common type of exam question. Make sure you adjust the limits of integration after substitution.

9780170459259

Question 7 (15 marks)

a i $\mu = p = 90\% = 0.9$ ✓

$$\sigma = \sqrt{\frac{p(1-p)}{n}} = \sqrt{\frac{0.9 \times 0.1}{100}} = 0.03 \; ✓$$

This is a straightforward question. Make sure you remember the formula for the standard deviation of the sample proportion because it is not on the HSC exam reference sheet.

ii 0.81 is 3 standard deviations below the mean:

$$\mu - 3\sigma = 0.9 - 3 \times 0.03 = 0.81 \; ✓$$

$$\therefore P(\hat{p} \le 0.81) \approx \frac{1}{2}(1 - 99.7\%)$$

$$= 0.15\% \; ✓$$

This is a straightforward question – use the HSC exam reference sheet for the empirical rule.

b Let $P(n)$ be the statement $(1 \times 2 \times 3) + (2 \times 3 \times 4) + \ldots + n(n+1)(n+2) = \frac{1}{4}n(n+1)(n+2)(n+3)$.

$P(1)$ is $1 \times 2 \times 3 = \frac{1}{4}(1)(1+1)(1+2)(1+3)$.

LHS $= 1 \times 2 \times 3$ RHS $= \frac{1}{4}(1)(1+1)(1+2)(1+3)$

   $= 6$ $= \frac{1}{4} \times 1 \times 2 \times 3 \times 4$

                                        $= 6$

                                        $=$ LHS

$\therefore P(1)$ is true. ✓

Assume $P(k)$ is true.

$P(k)$ is $(1 \times 2 \times 3) + (2 \times 3 \times 4) + \ldots + k(k+1)(k+2) = \frac{1}{4}k(k+1)(k+2)(k+3)$.   [*]

RTP $P(k+1)$: $(1 \times 2 \times 3) + (2 \times 3 \times 4) + \ldots + k(k+1)(k+2) + (k+1)(k+2)(k+3)$

$= \frac{1}{4}(k+1)(k+2)(k+3)(k+4)$.

LHS $= \frac{1}{4}k(k+1)(k+2)(k+3) + (k+1)(k+2)(k+3)$ by [*] ✓

   $= (k+1)(k+2)(k+3)\left[\frac{1}{4}k + 1\right]$

   $= (k+1)(k+2)(k+3)\frac{1}{4}(k+4)$

   $= \frac{1}{4}(k+1)(k+2)(k+3)(k+4)$

   $=$ RHS

$\therefore P(k+1)$ is true.

So by mathematical induction, $P(n)$ is true for all integers $n \ge 1$. ✓

This is a straightforward and common induction series proof.

c   Let the zeroes of $P(x)$ be $\alpha, -\alpha, \beta$ and γ.

From Viete's formulas (for sums and products of roots):

$$\alpha - \alpha + \beta + \gamma = \beta + \gamma = -\frac{-1}{1} = 1 \qquad [1]$$

$$\alpha(-\alpha) + \alpha\beta + \alpha\gamma - \alpha\beta - \alpha\gamma + \beta\gamma = -\alpha^2 + \beta\gamma = \frac{k}{1} = k \qquad [2]$$

$$\alpha(-\alpha)\beta + \alpha(-\alpha)\gamma + \alpha\beta\gamma - \alpha\beta\gamma = -\alpha^2(\beta + \gamma) = -\frac{25}{1} = -25 \quad [3]$$

$$\alpha(-\alpha)\beta\gamma = -\alpha^2\beta\gamma = \frac{300}{1} = 300 \qquad [4] \checkmark$$

Therefore, from [1] and [3], $-\alpha^2 = -25$ ✓

and thus, from [4], $\beta\gamma = -12$. ✓

Hence, from [2], $k = -25 + (-12) = -37$. ✓

By writing Viete's formulas before doing anything else, this question becomes quite straightforward.

d $\underset{\sim}{r} = \begin{pmatrix} ut \\ -5t^2 \end{pmatrix} \rightarrow \underset{\sim}{v} = \begin{pmatrix} u \\ -10t \end{pmatrix}$ ✓

Because the car lands at 45° to the horizontal, the gradient of the trajectory at impact is −1,

so $-\dfrac{10t}{u} = -1$

$\quad u = 10t.$ ✓

At that point:

$x = ut = 90$

So $t = \dfrac{90}{u}$. ✓

Thus, $u = 10\left(\dfrac{90}{u}\right)$

$\quad u^2 = 900$

$\quad u = 30\,\text{m/s}.$ ✓

Remember that the direction in which an object is moving depends on its *velocity*, not its position.
This final question is a complex 4-mark question with no scaffolding (clues).

Mathematics Extension 1

PRACTICE HSC EXAM 1

General instructions
- Reading time: 10 minutes
- Working time: 2 hours
- A reference sheet is provided on page 155 at the back of this book
- For questions in Section II, show relevant mathematical reasoning and/or calculations.

Total marks: 70

Section I – 10 questions, 10 marks
- Attempt Questions 1–10
- Allow about 15 minutes for this section

Section II – 4 questions, 60 marks
- Attempt Questions 11–14
- Allow about 1 hour 45 minutes for this section

Section I

10 marks
Attempt Questions 1–10
Allow about 15 minutes for this section

Circle the correct answer.

Question 1

Given $f(x) = \sqrt{\ln x}$, what is the domain of $f(x)$?

A $x \geq 0$

B $x > 0$

C $x \geq 1$

D $x > 1$

Question 2

What is the angle between the vectors $\begin{pmatrix} 24 \\ -7 \end{pmatrix}$ and $\begin{pmatrix} 12 \\ 16 \end{pmatrix}$?

A $\cos^{-1}(-0.352)$

B $\cos^{-1}(-0.048)$

C $\cos^{-1}(0.048)$

D $\cos^{-1}(0.352)$

Question 3

Find the derivative of $x \cos^{-1}\left(\dfrac{1}{x}\right)$.

A $\cos^{-1}\left(\dfrac{1}{x}\right) - \dfrac{1}{\sqrt{x^2 - 1}}$

B $\cos^{-1}\left(\dfrac{1}{x}\right) + \dfrac{1}{\sqrt{x^2 - 1}}$

C $\cos^{-1}\left(\dfrac{1}{x}\right) - \dfrac{1}{x\sqrt{x^2 - 1}}$

D $\cos^{-1}\left(\dfrac{1}{x}\right) + \dfrac{1}{x\sqrt{x^2 - 1}}$

Question 4

This diagram shows the graph of $y = f(x)$.

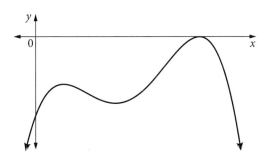

Which equation best describes this graph?

A $y = -(x - 4)^2(x^2 - x - 1)$

B $y = -(x - 4)^2(x^2 - x + 1)$

C $y = (x - 4)^2(x^2 - x - 1)$

D $y = (x - 4)^2(x^2 - x + 1)$

Question 5

How many ways can the letters of CENGAGE be arranged with alternating vowels and consonants?

A $\dfrac{7!}{2!}$

B $\dfrac{7!}{2!2!}$

C $\dfrac{4!3!}{2!}$

D $\dfrac{4!3!}{2!2!}$

Question 6

It is given that $\cos\theta = -\dfrac{\sqrt{5}}{3}$, where $\dfrac{\pi}{2} < \theta < \pi$.

What is the value of $\sin 2\theta$?

A $-\dfrac{4\sqrt{5}}{9}$

B $-\dfrac{4}{9}$

C $\dfrac{4}{9}$

D $\dfrac{4\sqrt{5}}{9}$

Question 7

Which is the slope field for the differential equation $\dfrac{dy}{dx} = \dfrac{1}{x} - \dfrac{1}{y}$?

A

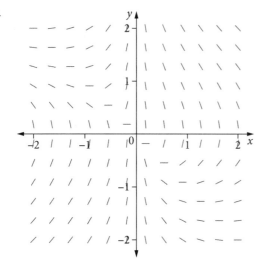

B

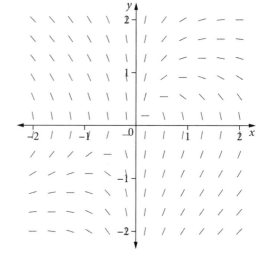

C

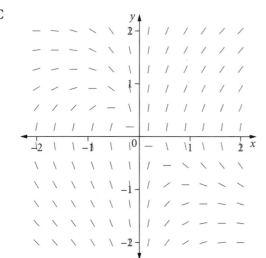

D

Question 8

Find the value of $\dbinom{n}{0} - \dbinom{n}{1} + \dbinom{n}{2} - \dbinom{n}{3} + \cdots + (-1)^n \dbinom{n}{n}$.

A $(-1)^n$

B 0

C 1

D 2^n

Question 9

The surface area A of a cube is increasing at a constant rate k.

Which expression gives the rate of change of the volume V of the cube when the side length is s?

A $\dfrac{dV}{dt} = \dfrac{ks}{4}$

B $\dfrac{dV}{dt} = 3ks^2$

C $\dfrac{dV}{dt} = 12ks$

D $\dfrac{dV}{dt} = 36ks^3$

Question 10

Which of the following equations is false?

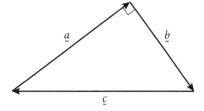

A $\text{proj}_{\underset{\sim}{b}}\underset{\sim}{a} = \underset{\sim}{c}$

B $\left|\underset{\sim}{a} + \underset{\sim}{b}\right| = \left|\underset{\sim}{c}\right|$

C $\underset{\sim}{a} + \underset{\sim}{b} + \underset{\sim}{c} = 0$

D $\underset{\sim}{a} \cdot \underset{\sim}{a} + \underset{\sim}{b} \cdot \underset{\sim}{b} = \underset{\sim}{c} \cdot \underset{\sim}{c}$

Section II

60 marks
Attempt Questions 11–14
Allow about 1 hour and 45 minutes for this section

- Answer the questions in the spaces provided. These spaces provide guidance for the expected length of response.
- Your responses should include relevant mathematical reasoning and/or calculations.

Question 11 (15 marks)

a Is $x - 43$ a factor of $x^2 + 4x - 2021$? Give a reason for your answer.   2 marks

b Find $\int 2\cos 2x \cos x \, dx$.   2 marks

c Solve $\dfrac{6}{x - 2} > 1$.   3 marks

Question 11 continues on page 103

Question 11 (continued)

d   The diagram below shows the graph of $y = f(x)$.   3 marks

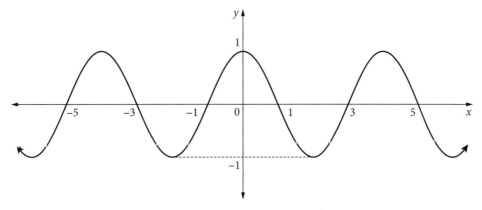

Sketch on the diagram the graph of $y^2 = f(x)$.

e   Find $\int 2\cos^2 2x\, dx$.   2 marks

f   Express $\sin x + \sin\left(x + \dfrac{\pi}{3}\right)$ in the form $A \sin(x + \alpha)$, where $A > 0$ and $0 < \alpha < \dfrac{\pi}{2}$.   3 marks

End of Question 11

Question 12 (15 marks)

a A coin is flipped 1000 times. Let X represent the number of tails obtained. 4 marks

By using a normal approximation, find the value of $P(484 < X < 532)$. Give your answer as a percentage correct to one decimal place.

b ©NESA 2001 HSC EXAM, QUESTION 3(c)

   i Starting from the identity $\sin(\theta + 2\theta) = \sin\theta\cos 2\theta + \cos\theta\sin 2\theta$, and using the double angle 2 marks formulae, prove the identity

$$\sin 3\theta = 3\sin\theta - 4\sin^3\theta.$$

Question 12 continues on page 105

Question 12 (continued)

ii Hence, solve the equation $\sin 3\theta = 2\sin\theta$ for $0 \le \theta \le 2\pi$.   3 marks

c ©NESA 2002 HSC EXAM, QUESTION 3(a)

Seven people are to be seated at a round table.

i How many seating arrangements are possible?   1 mark

ii Two people, Kevin and Jill, refuse to sit next to each other. How many seating arrangements are then possible?   2 marks

d Find the function that satisfies the differential equation $\dfrac{dy}{dx} = e^{xy}$, given that the graph of $y = f(x)$ goes through the origin.   3 marks

End of Question 12

Question 13 (15 marks)

a   Use the substitution $x = u^2$ to evaluate $\int_{\frac{1}{4}}^{\frac{3}{4}} \dfrac{dx}{\sqrt{x(1-x)}}$.   3 marks

b   Use the pigeonhole principle to prove that, in any set of at least 11 distinct integers, there exist 2 distinct integers whose difference is a multiple of 10.   3 marks

Question 13 continues on page 107

9780170459259

Question 13 (continued)

c   An object is launched from the origin at an angle of 30° to the horizontal, with an initial   4 marks
velocity of 60 m/s. The acceleration vector is $\underset{\sim}{a} = -10\underset{\sim}{j}$.

Find the displacement vector $\underset{\sim}{r}$ and hence, show that the path of the object has the equation

$$y = \frac{1}{\sqrt{3}}x - \frac{1}{54}x^2.$$

d   ©NESA   2002 HSC EXAM, QUESTION 5(c)

Consider the function

$$f(x) = 2\sin^{-1}\sqrt{x} - \sin^{-1}(2x - 1) \text{ for } 0 \le x \le 1.$$

i   Show that $f'(x) = 0$ for $0 < x < 1$.   3 marks

ii   Sketch the graph of $y = f(x)$.   2 marks

End of Question 13

Question 14 (15 marks)

a A simplified model for the number of deaths in a pandemic is the logistic equation $\dfrac{dD}{dt} = AD(B - D)$,

where D represents the number of deaths, t represents time (in days), and A and B are positive constants, where $B > D$.

    i Using the fact that $\dfrac{1}{D(B - D)} = \dfrac{1}{B}\left(\dfrac{1}{D} + \dfrac{1}{B - D}\right)$, show that the general solution to the above   3 marks

    equation is $D = \dfrac{B}{1 + Ce^{-ABt}}$, where C is a positive constant.

a

Question 14 continues on page 109

Question 14 (continued)

ii   Assume that the first death occurred when $t = 0$, and there are 800 deaths in total during   3 marks
the pandemic. When $t = 100$, there had been 101 deaths.

Based on this information, predict how many deaths occurred by $t = 200$.

Question 14 continues on page 110

Question 14 (continued)

b Prove by mathematical induction that $3 \times 2^{3n-2} + 1$ is divisible by 7, for all integers $n \geq 1$.   3 marks

b

Question 14 continues on page 111

Question 14 (continued)

c   The triangle *OAB* is defined by the vectors $\underset{\sim}{a} = \overrightarrow{OA}$ and $\underset{\sim}{b} = \overrightarrow{OB}$, as shown below.

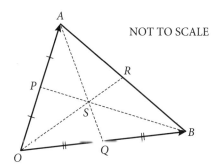

NOT TO SCALE

P and *Q* are the midpoints of *OA* and *OB* respectively. *AQ* and *BP* intersect at *S*.
OS is extended to meet *AB* at *R*.

i   Show that $\overrightarrow{OS} = \dfrac{1}{2}\underset{\sim}{b} + m\left(\underset{\sim}{a} - \dfrac{1}{2}\underset{\sim}{b}\right)$ for some constant *m*.   2 marks

ii   Given that $\overrightarrow{OS} = \dfrac{1}{2}\underset{\sim}{a} + n\left(\underset{\sim}{b} - \dfrac{1}{2}\underset{\sim}{a}\right)$ also for some constant *n*, show that $\overrightarrow{OS} = \dfrac{1}{3}(\underset{\sim}{a} + \underset{\sim}{b})$.   2 marks

iii   Hence, show that $\overrightarrow{OR} = \dfrac{1}{2}(\underset{\sim}{a} + \underset{\sim}{b})$.   2 marks

END OF PAPER

SECTION II EXTRA WRITING SPACE

WORKED SOLUTIONS

Section I (1 mark each)

Question 1

C $\ln x \geq 0$, so
$$x \geq 1$$

> This is a Year 11 understanding question. Recall that $\ln x < 0$ for $x < 1$ restricts the domain for $f(x)$ due to the square root.

Question 2

D $\begin{pmatrix} 24 \\ -7 \end{pmatrix} \cdot \begin{pmatrix} 12 \\ 16 \end{pmatrix} = \left| \begin{pmatrix} 24 \\ -7 \end{pmatrix} \right| \left| \begin{pmatrix} 12 \\ 16 \end{pmatrix} \right| \cos \theta$

$288 - 112 = 25 \times 20 \times \cos \theta$

So $\cos \theta = \dfrac{176}{500} = 0.352$.

> This is a straightforward vector question.

Question 3

B $\dfrac{d}{dx}\left(x \cos^{-1} \dfrac{1}{x} \right) = \cos^{-1}\left(\dfrac{1}{x}\right) + x \dfrac{\frac{1}{x^2}}{\sqrt{1 - \left(\frac{1}{x}\right)^2}}$

$= \cos^{-1}\left(\dfrac{1}{x}\right) + \dfrac{1}{\sqrt{x^2 - 1}}$

> Use the HSC exam reference sheet.

Question 4

B From the graph, the leading coefficient is negative (options A or B). It has a double root and 2 non-real roots ($x^2 - x + 1$ in option B has no real roots).

OR

Substituting $x = 0$ into options A and B, only B gives a negative y-intercept.

> This is a Year 11 question on understanding shapes of polynomials.

Question 5

D CVCVCVC $\rightarrow \dfrac{4!}{2!} \times \dfrac{3!}{2!} = \dfrac{4!3!}{2!2!}$

> This is a very common type of Year 11 permutations question. Recall that since we have 2 identical consonants (G) and vowels (E), we divide their corresponding combinations by 2! to 'uncount' combinations where they are swapped.

Question 6

A $\sin 2\theta = 2 \sin \theta \cos \theta$

$\sin \theta = \sqrt{1 - \cos^2 \theta} \quad \left(\dfrac{\pi}{2} < \theta < \pi \right)$

$= \sqrt{1 - \dfrac{5}{9}}$

$= \dfrac{2}{3}$

So $\sin 2\theta = 2 \times \dfrac{2}{3} \times \left(-\dfrac{\sqrt{5}}{3} \right)$

$= -\dfrac{4\sqrt{5}}{9}$.

> Practise the trigonometric identities.

Question 7

B If $x = y$, then

$\dfrac{dy}{dx} = 0$, so flat on the line $y = x$ and $(1, 1)$, $(2, 2)$:   (options B or D)

If $x > 0, y < 0$

$\dfrac{dy}{dx} > 0$, so positive gradients in 4th quadrant
          (option B)

> Look for features that let you eliminate options.

Question 8

B Test using values in Pascal's triangle:

$$
\begin{array}{ccccccccc}
& & & 1 & & & & & \\
& & 1 & & 1 & & 1-1 & & = 0 \\
& 1 & & 2 & & 1 & 1-2+1 & & = 0 \\
1 & & 3 & & 3 & & 1 \quad 1-3+3-1 & & = 0 \\
1 & 4 & & 6 & & 4 & 1 \quad 1-4+6-4+1 & & = 0
\end{array}
$$

OR

Formal proof:

$$(1+x)^n = \binom{n}{0}1^n x^0 + \binom{n}{1}1^{n-1}x^1 + \cdots + \binom{n}{n-1}1^1 x^{n-1} + \binom{n}{n}1^0 x^n$$

$$= \binom{n}{0}x^0 + \binom{n}{1}x^1 + \cdots + \binom{n}{n-1}x^{n-1} + \binom{n}{n}x^n$$

For alternating signs in the given sum, substitute $x = -1$:

$$(1+[-1])^n = \binom{n}{0}(-1)^0 + \binom{n}{1}(-1)^1 + \cdots + \binom{n}{n-1}(-1)^{n-1} + \binom{n}{n}(-1)^n$$

$$0^n = \binom{n}{0} - \binom{n}{1} + \cdots + \binom{n}{n-1}(-1)^{n-1} + \binom{n}{n}(-1)^n$$

$$\therefore \binom{n}{0} - \binom{n}{1} + \binom{n}{2} - \binom{n}{3} + \cdots + \binom{n}{n-1}(-1)^{n-1} + \binom{n}{n}(-1)^n = 0$$

This is a tricky Year 11 question on understanding the binomial theorem.

Question 9

A $\dfrac{dA}{dt} = k,\ A = 6s^2,\ V = s^3$

$$\frac{dV}{dt} = \frac{dV}{ds} \times \frac{ds}{dA} \times \frac{dA}{dt}$$

$$= 3s^2 \times \frac{1}{12s} \times k$$

$$= \frac{sk}{4}$$

It is important to read the questions very carefully here. After that, it is a straightforward application of related rates.

Question 10

A $\underset{\sim}{c}$ is not the projection of $\underset{\sim}{a}$ on $\underset{\sim}{b}$ because $\underset{\sim}{a}$ is perpendicular to $\underset{\sim}{b}$.

B and C are true because $\underset{\sim}{a} + \underset{\sim}{b} = -\underset{\sim}{c}$ and D is true because of Pythagoras' theorem.

WORKED SOLUTIONS

Section II ($\checkmark$ = 1 mark)

Question 11 (15 marks)

a $P(43) = 43^2 - 4 \times 43 - 2021 = 0$ $\checkmark$

Therefore, $(x - 43)$ is a factor because of the factor theorem. $\checkmark$

This is a Year 11 understanding question on polynomials. Make sure you show working and give a reason.

b From the reference sheet:

$$\cos A \cos B = \frac{1}{2}\left[\cos(A - B) + \cos(A - B)\right]$$

Let $A = 2x$, $B = x$:

$$\cos 2x \cos x = \frac{1}{2}\left[\cos(2x - x) + \cos(2x + x)\right]$$
$$2\cos 2x \cos x = \cos x + \cos 3x \quad \checkmark$$

$$\int 2\cos 2x \cos x \, dx = \int (\cos x + \cos 3x) \, dx$$
$$= \sin x + \frac{1}{3}\sin 3x + c \quad \checkmark$$

Use the HSC exam reference sheet and don't forget the constant term '+ c'.

c $$\frac{6}{x - 2} > 1$$
$$6(x - 2) > (x - 2)^2$$
$$6x - 12 > x^2 - 4x + 4$$
$$x^2 - 10x + 16 < 0 \quad \checkmark$$
$$(x - 8)(x - 2) < 0 \quad \checkmark$$

So $2 < x < 8$. $\checkmark$

To solve this inequality, we multiply both sides by $(x - 2)^2$ instead of $(x - 2)$ because $(x - 2)^2$ is always positive and won't reverse the inequality symbol. Draw a sketch to ensure that the inequality in the answer is correct.

d

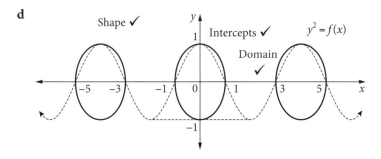

This is a common Year 11 graphing question. Be very clear when sketching and include the original function. Make sure to use half a page.

e $\int 2\cos^2 2x \, dx = \int (1 + \cos 4x) \, dx$ $\checkmark$ (from the reference sheet)

$$= x + \frac{1}{4}\sin 4x + c \quad \checkmark$$

This is another integration involving the square of sine or cosine – use the HSC exam reference sheet.

f $\sin x + \sin\left(x + \dfrac{\pi}{3}\right) = \sin x + \sin x \cos\dfrac{\pi}{3} + \cos x \sin\dfrac{\pi}{3}$

$$= \sin x + \dfrac{1}{2}\sin x + \dfrac{\sqrt{3}}{2}\cos x$$

$$= \dfrac{3}{2}\sin x + \dfrac{\sqrt{3}}{2}\cos x \checkmark$$

Let $\dfrac{3}{2}\sin x + \dfrac{\sqrt{3}}{2}\cos x = A\sin(x + \alpha)$.

$$= A(\sin x \cos\alpha + \cos x \sin\alpha)$$

$$= A\sin x \cos\alpha + A\cos x \sin\alpha$$

$\therefore A\cos\alpha = \dfrac{3}{2}$ [1]

$\therefore \sin\alpha = \dfrac{\sqrt{3}}{2}$ [2]

[2] ÷ [1]: $\tan\alpha = \dfrac{\frac{\sqrt{3}}{2}}{\frac{3}{2}} = \dfrac{\sqrt{3}}{3} = \dfrac{1}{\sqrt{3}}$

$$\alpha = \dfrac{\pi}{6} \checkmark$$

$[1]^2 + [2]^2$: $A^2\cos^2\alpha + A^2\sin^2\alpha = \left(\dfrac{3}{2}\right)^2 + \left(\dfrac{\sqrt{3}}{2}\right)^2$

$$A^2(\cos^2\alpha + \sin^2\alpha) = \dfrac{12}{4} = 3$$

$$A = \sqrt{3}$$

$\therefore \sin x + \sin\left(x + \dfrac{\pi}{3}\right) = \sqrt{3}\sin\left(x + \dfrac{\pi}{6}\right) \checkmark$

OR

Using the formula

$$\dfrac{1}{2}\left[\sin(A + B) + \sin(A - B)\right] = \sin A \cos B$$

with $A = x + \dfrac{\pi}{6}$, $B = \dfrac{\pi}{6}$.

$$\dfrac{1}{2}\left[\sin\left(x + \dfrac{\pi}{6} + \dfrac{\pi}{6}\right) + \sin\left(x + \dfrac{\pi}{6} - \dfrac{\pi}{6}\right)\right] = \sin\left(x + \dfrac{\pi}{6}\right)\cos\dfrac{\pi}{6} \checkmark$$

$$\dfrac{1}{2}\left[\sin\left(x + \dfrac{\pi}{3}\right) + \sin x\right] = \dfrac{\sqrt{3}}{2}\sin\left(x + \dfrac{\pi}{6}\right) \checkmark$$

$$\sin x + \sin\left(x + \dfrac{\pi}{3}\right) = \sqrt{3}\sin\left(x + \dfrac{\pi}{6}\right) \checkmark$$

This is a variation on a common type of question: the auxiliary angle method.

Question 12 (15 marks)

a $n = 1000, p = \dfrac{1}{2}$ ✓

$\mu = 1000 \times \dfrac{1}{2}$
$= 500$

$\sigma = \sqrt{1000 \times \dfrac{1}{2} \times \dfrac{1}{2}}$
≈ 15.81 ✓

404 is 2 standard deviations below the mean, 500, and 532 is 1 standard deviation above.

$P(484 < X < 532) \approx P(-1 < z < 2)$ ✓

$\approx \dfrac{1}{2}68\% + \dfrac{1}{2}95\%$

$= 81.5\%$ ✓

This is a common sample proportions question, based on understanding and using the HSC exam reference sheet.

b **i** $\sin 3\theta = \sin(2\theta + \theta)$
$= \sin 2\theta \cos \theta + \cos 2\theta \sin \theta$
$= (2 \sin \theta \cos \theta) \cos \theta + (\cos^2 \theta - \sin^2 \theta) \sin \theta$
$= 2 \sin \theta \cos^2 \theta + \sin \theta \cos^2 \theta - \sin^3 \theta$
$= 3 \sin \theta \cos^2 \theta - \sin^3 \theta$ ✓
$= 3 \sin \theta(1 - \sin^2 \theta) - \sin^3 \theta$
$= 3 \sin \theta - 3 \sin^3 \theta - \sin^3 \theta$
$= 3 \sin \theta - 4 \sin^3 \theta$ ✓

ii $\sin 3\theta = 2 \sin \theta$
$3 \sin \theta - 4 \sin^3 \theta = 2 \sin \theta$
$4 \sin^3 \theta - \sin \theta = 0$ ✓
$\sin \theta(4 \sin^2 \theta - 1) = 0$

$\sin \theta = 0$ or $\sin^2 \theta = \dfrac{1}{4}$

$\sin \theta = 0$ or $\sin \theta = \pm\dfrac{1}{2}$ ✓

$\theta = 0, \pi, 2\pi$ or $\dfrac{\pi}{6}, \dfrac{5\pi}{6}, \dfrac{7\pi}{6}, \dfrac{11\pi}{6}$ ✓

$\theta = 0, \dfrac{\pi}{6}, \dfrac{5\pi}{6}, \pi, \dfrac{7\pi}{6}, \dfrac{11\pi}{6}, 2\pi$

This is a typical trigonometric equation question with factorising from the 2001 HSC exam. Make sure to include ALL the solutions in the domain. Only $\frac{1}{3}$ of the students in 2001 listed the 7 solutions.

c **i** $(7 - 1)! = 720$ seating arrangements. ✓

This 2002 HSC question tests the Year 11 permutations topic. Many students still have trouble with arrangements in a circle. Learn this correctly.

ii Kevin and Jill sitting next to each other:

$2 \times 5! = 240$ seating arrangements. ✓

So $720 - 240 = 480$ seating arrangements. ✓

This is a typical Year 11 combinatorics complement question; that is, find the opposite and subtract it from the total.

d $\dfrac{dy}{dx} = e^{xy}$

$\int e^{-y} \, dy = \int e^x \, dx$

Thus, $-e^{-y} = e^x + c.$ ✓

When $x = 0, y = 0$:

$-e^0 = e^0 + c$
$-1 = 1 + c$
$-2 = c$

Hence, $-e^{-y} = e^x - 2$ ✓
$e^{-y} = 2 - e^x.$

So $y = -\ln(2 - e^x).$ ✓

This is a straightforward differential equation requiring separation of variables. Don't forget the constant term '+ c'. The constant after integration can go on either side, and you'll still get the same result.

Question 13 (15 marks)

a $x = u^2$

$dx = 2u \, du$ ✓

When $x = \dfrac{1}{4}, u = \dfrac{1}{2}.$

When $x = \dfrac{3}{4}, u = \dfrac{\sqrt{3}}{2}.$

So $\displaystyle\int_{\frac{1}{4}}^{\frac{3}{4}} \dfrac{dx}{\sqrt{x(1-x)}} = \int_{\frac{1}{2}}^{\frac{\sqrt{3}}{2}} \dfrac{2u\,du}{\sqrt{u^2(1-u^2)}}$

$\displaystyle = 2\int_{\frac{1}{2}}^{\frac{\sqrt{3}}{2}} \dfrac{du}{\sqrt{1-u^2}}$ ✓

$= 2\left[\sin^{-1} u\right]_{\frac{1}{2}}^{\frac{\sqrt{3}}{2}}$

$= 2\left[\sin^{-1}\dfrac{\sqrt{3}}{2} - \sin^{-1}\dfrac{1}{2}\right]$

$= 2\left(\dfrac{\pi}{3} - \dfrac{\pi}{6}\right)$

$= \dfrac{\pi}{3}.$ ✓

Make sure you change the limits of integration.

WORKED SOLUTIONS

b Let the pigeonholes be the remainder when each integer is divided by 10. Then there are 10 pigeonholes (0 ... 9). ✓

By the pigeonhole principle, 2 integers must have the same remainder when divided by 10; let it be r. ✓

Then the 2 integers are of the form $10p + r$ and $10q + r$. But the difference is $(10p + r) - (10q + r)$ $= 10p - 10q = 10(p - q)$, which is a multiple of 10. ✓

> The important step is to identify the pigeonholes. Once you have done that, the question should be straightforward. Revise the pigeonhole principle from Year 11 and then practise as many of these questions as possible.

c $\underset{\sim}{a} = -10\underset{\sim}{j} \rightarrow \underset{\sim}{v} = -10t\underset{\sim}{j} + \underset{\sim}{c_1}$

$\underset{\sim}{v}(0) = 60\cos 30°\underset{\sim}{i} + 60\sin 30°\underset{\sim}{j} = 30\sqrt{3}\underset{\sim}{i} + 30\underset{\sim}{j} = \underset{\sim}{c_1}$

$\therefore \underset{\sim}{v} = 30\sqrt{3}\underset{\sim}{i} + (30 - 10t)\underset{\sim}{j}$ ✓

Hence, $\underset{\sim}{r} = 30(\sqrt{3})t\underset{\sim}{i} + (30t - 5t^2)\underset{\sim}{j} + \underset{\sim}{c_2}$

$\underset{\sim}{r}(0) = \underset{\sim}{0} = \underset{\sim}{c_2}.$

Hence, $\underset{\sim}{r} = 30(\sqrt{3})t\underset{\sim}{i} + (30t - 5t^2)\underset{\sim}{j}.$ ✓

Thus, $x = 30(\sqrt{3})t \rightarrow t = \dfrac{x}{30\sqrt{3}}.$ ✓

So $y = 30t - 5t^2 = \dfrac{1}{\sqrt{3}}x - 5\left(\dfrac{x^2}{2700}\right)$

$= \dfrac{1}{\sqrt{3}}x - \dfrac{1}{540}x^2.$ ✓

> This is a common projectile motion question, and should be one of the types practised often.

d i $\dfrac{d}{dx}2\sin^{-1}\sqrt{x} = 2\dfrac{\frac{1}{2\sqrt{x}}}{\sqrt{1-x}} = \dfrac{1}{\sqrt{x(1-x)}}$ $(0 < x < 1)$ ✓

$\dfrac{d}{dx}\sin^{-1}(2x - 1) = \dfrac{2}{\sqrt{1 - (2x - 1)^2}}$

$= \dfrac{2}{\sqrt{1 - 4x^2 + 4x - 1}}$

$= \dfrac{2}{\sqrt{-4x^2 + 4x}}$

$= \dfrac{1}{\sqrt{x(1-x)}}$ $(0 < x < 1)$ ✓

So $\dfrac{d}{dx}\left(2\sin^{-1}\sqrt{x} - \sin^{-1}(2x - 1)\right) = 0.$ $(0 < x < 1)$ ✓

> This is quite a challenging chain rule differentiation involving inverse sine functions, from the 2002 HSC exam. Many students wasted time writing pages of incorrect algebra for this 3-mark question. Although the solution is complex, there aren't that many steps.

ii As $f'(x) = 0$, $f(x) = c$, a constant. Substitute a convenient value of x to find c.

$$x = 0 \rightarrow 2\sin^{-1}\sqrt{x} - \sin^{-1}(2x - 1)$$
$$= 2(0) - \sin^{-1}(-1)$$
$$= -\left(-\frac{\pi}{2}\right)$$
$$= \frac{\pi}{2} \checkmark$$

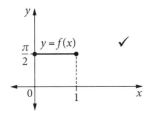

Students needed to realise that if $f'(x) = 0$ in part **i**, then $f(x)$ is a constant whose graph is a horizontal line. You also must restrict the domain of the line to [0, 1] to score both marks.

Question 14 (15 marks)

a i $\dfrac{dD}{dt} = AD(B - D)$

$$\frac{dt}{dD} = \frac{1}{AD(B - D)} = \frac{1}{AB}\left(\frac{1}{D} + \frac{1}{B - D}\right)$$

Hence, $t = \dfrac{1}{AB}(\ln D - \ln(B - D)) + c$

$$= \frac{1}{AB}\ln\left(\frac{D}{B - D}\right) + c. \checkmark$$

Thus, $\ln\left(\dfrac{D}{B - D}\right) = ABt - c$

$$\frac{D}{B - D} = e^{ABt - c}. \checkmark$$

$D = Be^{ABt-c} - De^{ABt-c}$

Hence, $D(1 + e^{ABt-c}) = Be^{ABt-c}$

$$\therefore D = \frac{Be^{ABt-c}}{1 + e^{ABt-c}}$$
$$= \frac{B}{1 + e^{-ABt+c}}$$
$$= \frac{B}{1 + Ce^{-ABt}}, \text{ where } C = e^c. \checkmark$$

The logistic equation is complex but common. Learn and practise the processes. Don't forget the constant term '+ c'.

ii $\lim\limits_{t \to \infty} D = B = 800$

When $t = 0$, $D = 1$:

$$1 = \frac{800}{1 + Ce^{-800A(0)}}$$
$$1 = \frac{800}{1 + C}$$
$$C = 799 \checkmark$$
$$D = \frac{800}{1 + 799e^{-800At}}$$

When $t = 100$, $D = 101$:

$$101 = \frac{800}{1 + 799e^{-800A(100)}}$$
$$e^{-80\,000A} = \frac{699}{80\,699}$$
$$A = \frac{1}{80\,000}\ln\left(\frac{80\,699}{699}\right)$$
$$\approx 0.000\,059\,36 \checkmark$$

$\therefore$ When $t = 200$:

$$D = \frac{800}{1 + 799e^{-800A(200)}}$$
$$\approx 754.755$$

There were about 755 deaths after 200 days. $\checkmark$

This question involves a lot of algebra and calculation that you need to be careful with.

b Let $P(n)$ be the statement $3 \times 2^{3n-2} + 1$ is divisible by 7.

$P(1)$ is $3 \times 2^{3(1)-2} + 1 = 7$, which is divisible by 7.

$\therefore P(1)$ is true. $\checkmark$

Assume $P(k)$ is true.

Thus $3 \times 2^{3k-2} + 1 = 7p$ for some integer p.

$\therefore 3(2^{3k-2}) = 7p - 1$ [*]

RTP $P(k + 1)$: $3 \times 3^{(k+1)-2} + 1$ is divisible by 7.

$3 \times 2^{3(k+1)-2} + 1$
$= 3 \times 2^{3k+1} + 1$
$= 3 \times 2^{3k-2+3} + 1$
$= 3 \times 2^{3k-2}2^3 + 1$
$= (7p - 1)8 + 1$ from [*] $\checkmark$
$= 56p - 8 + 1$
$= 56p - 7$
$= 7(8p - 1)$, which is divisible by 7.

$\therefore P(k + 1)$ is true.

So by mathematical induction, $P(n)$ is true for all integers $n \geq 1$. $\checkmark$

This is a typical divisibility induction question.

c i $\overrightarrow{OQ} = \dfrac{1}{2}\overrightarrow{OB} = \dfrac{1}{2}\underset{\sim}{b}$

Hence, $\overrightarrow{QA} = \overrightarrow{OA} - \overrightarrow{OQ}$

$$= \underset{\sim}{a} - \dfrac{1}{2}\underset{\sim}{b} \quad\checkmark$$

$\overrightarrow{OS} = \overrightarrow{OQ} + \overrightarrow{QS}$

$\qquad = \overrightarrow{OQ} + m\overrightarrow{QA}$ for some constant m, $0 < m < 1$

$$= \dfrac{1}{2}\underset{\sim}{b} + m\left(\underset{\sim}{a} - \dfrac{1}{2}\underset{\sim}{b}\right) \quad\checkmark$$

> You need a good visual understanding of vectors for this challenging final question.

ii $\overrightarrow{OS} = \dfrac{1}{2}\underset{\sim}{b} + m\left(\underset{\sim}{a} - \dfrac{1}{2}\underset{\sim}{b}\right)$ (from part **i**)

Given $\overrightarrow{OS} = \dfrac{1}{2}\underset{\sim}{a} + n\left(\underset{\sim}{b} - \dfrac{1}{2}\underset{\sim}{a}\right)$:

$$\therefore \dfrac{1}{2}\underset{\sim}{b} + m\left(\underset{\sim}{a} - \dfrac{1}{2}\underset{\sim}{b}\right) = \dfrac{1}{2}\underset{\sim}{a} + n\left(\underset{\sim}{b} - \dfrac{1}{2}\underset{\sim}{a}\right)$$

$$\dfrac{1}{2}\underset{\sim}{b} + m\underset{\sim}{a} - \dfrac{1}{2}m\underset{\sim}{b} = \dfrac{1}{2}\underset{\sim}{a} + n\underset{\sim}{b} - \dfrac{1}{2}n\underset{\sim}{a}$$

$$\dfrac{1}{2}(1 - m)\underset{\sim}{b} + m\underset{\sim}{a} = \dfrac{1}{2}(1 - n)\underset{\sim}{a} + n\underset{\sim}{b}$$

Equating coefficients:

$$\dfrac{1}{2}(1 - m) = n \qquad [1]$$

$$m = \dfrac{1}{2}(1 - n) \quad [2] \quad\checkmark$$

Substitute [1] into [2]:

$$m = \dfrac{1}{2}\left(1 - \dfrac{1}{2}(1 - m)\right)$$

$$2m = 1 - \dfrac{1}{2}(1 - m)$$

$$4m = 2 - (1 - m)$$

$$4m = 1 + m$$

$$3m = 1$$

$$m = \dfrac{1}{3}$$

Substitute into [1]:

$$\dfrac{1}{2}\left(1 - \dfrac{1}{3}\right) = n$$

$$n = \dfrac{1}{2} \times \dfrac{2}{3} = \dfrac{1}{3}$$

$\therefore$ Equation of $\overrightarrow{OS}$ is:

$$\overrightarrow{OS} = \dfrac{1}{2}\underset{\sim}{b} + \dfrac{1}{3}\left(\underset{\sim}{a} - \dfrac{1}{2}\underset{\sim}{b}\right)$$

$$= \dfrac{1}{2}\underset{\sim}{b} + \dfrac{1}{3}\underset{\sim}{a} - \dfrac{1}{6}\underset{\sim}{b}$$

$$= \dfrac{1}{3}\underset{\sim}{a} + \dfrac{1}{3}\underset{\sim}{b}$$

$$= \dfrac{1}{3}(\underset{\sim}{a} + \underset{\sim}{b}) \quad\checkmark$$

> Note the link to part **i** and the need to solve simultaneous equations.

iii $\overrightarrow{OR} = \overrightarrow{OA} + p\overrightarrow{AB}$ $\qquad$ for some constant p,
$\qquad = \underset{\sim}{a} + p(\underset{\sim}{b} - \underset{\sim}{a})$ $\qquad 0 < p < 1$

$\overrightarrow{OR} = q\overrightarrow{OS}$ for some constant q, $q > 1$

$$= \dfrac{q}{3}(\underset{\sim}{a} + \underset{\sim}{b}) \quad \text{(from part \textbf{ii})}$$

$$\therefore \underset{\sim}{a} + p(\underset{\sim}{b} - \underset{\sim}{a}) = \dfrac{q}{3}(\underset{\sim}{a} + \underset{\sim}{b})$$

$$\underset{\sim}{a} + p\underset{\sim}{b} - p\underset{\sim}{a} = \dfrac{q}{3}\underset{\sim}{a} + \dfrac{q}{3}\underset{\sim}{b}$$

$$(1 - p)\underset{\sim}{a} + p\underset{\sim}{b} = \dfrac{q}{3}\underset{\sim}{a} + \dfrac{q}{3}\underset{\sim}{b}$$

Equating coefficients:

$$1 - p = \dfrac{q}{3} \quad [1]$$

$$p = \dfrac{q}{3} \quad [2] \quad\checkmark$$

[1] − [2]:

$$1 - 2p = 0$$

$$1 = 2p$$

$$p = \dfrac{1}{2}$$

Substitute into [2]:

$$\dfrac{1}{2} = \dfrac{q}{3}$$

$$q = \dfrac{3}{2}$$

$$\therefore \overrightarrow{OR} = \dfrac{q}{3}(\underset{\sim}{a} + \underset{\sim}{b})$$

$$= \dfrac{\frac{3}{2}}{3}(\underset{\sim}{a} + \underset{\sim}{b})$$

$$= \dfrac{1}{2}(\underset{\sim}{a} + \underset{\sim}{b}) \quad\checkmark$$

> This is similar to parts **i** and **ii**. Intersection questions should always use the intersecting vectors in the answer. Other approaches tend to lead to dead ends.

9780170459259

Mathematics Extension 1

PRACTICE HSC EXAM 2

General instructions	• Reading time: 10 minutes
	• Working time: 2 hours
	• A reference sheet is provided on page 155 at the back of this book
	• For questions in Section II, show relevant mathematical reasoning and/or calculations

Total marks: 70	**Section I – 10 questions, 10 marks**
	• Attempt Questions 1–10
	• Allow about 15 minutes for this section
	Section II – 4 questions, 60 marks
	• Attempt Questions 11–14
	• Allow about 1 hours and 45 minutes for this section

Section I

10 marks
Attempt Questions 1–10
Allow about 15 minutes for this section

Circle the correct answer.

Question 1

What is the domain of the function $f(x) = \ln(\sqrt{x})$?

A $x \geq 0$

B $x > 0$

C $x \geq 1$

D $x > 1$

Question 2

What is the sum of the roots of $P(x) = x^3 - 3x^2 - 2025x - 2021$?

A -2025

B -1

C 3

D 2025

Question 3

Which expression is equal to $\int \cos^2 4x \, dx$?

A $\dfrac{1}{2}x - \dfrac{1}{16}\cos 8x + c$

B $\dfrac{1}{2}x + \dfrac{1}{16}\cos 8x + c$

C $\dfrac{1}{2}x - \dfrac{1}{16}\sin 8x + c$

D $\dfrac{1}{2}x + \dfrac{1}{16}\sin 8x + c$

Question 4

For what value of k are the vectors $\begin{pmatrix} 2 \\ k \end{pmatrix}$ and $\begin{pmatrix} -1 \\ k-1 \end{pmatrix}$ perpendicular?

A $k = -2$

B $k = -1$

C $k = 0$

D $k = 1$

Question 5

Which graph best represents $y = \dfrac{\pi}{2} - \cos^{-1}(-x)$?

A

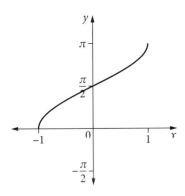

B

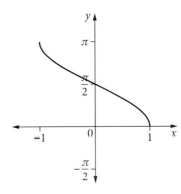

C

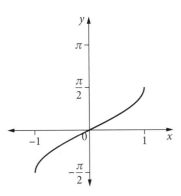

D

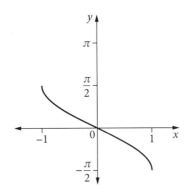

Question 6

What is the general solution to the equation $4\sin^2 x = 4\sin x + 3$, if n is an integer?

A $x = \dfrac{\pi}{6} + 2n\pi, \dfrac{5\pi}{6} + 2n\pi$

B $x = \dfrac{7\pi}{6} + 2n\pi, \dfrac{11\pi}{6} + 2n\pi$

C $x = \dfrac{\pi}{3} + 2n\pi, \dfrac{2\pi}{3} + 2n\pi$

D $x = \dfrac{4\pi}{3} + 2n\pi, \dfrac{5\pi}{3} + 2n\pi$

Question 7

Which differential equation has the slope field shown?

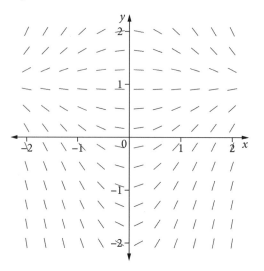

A $\dfrac{dy}{dx} = x(1 - y)$

B $\dfrac{dy}{dx} = x(y - 1)$

C $\dfrac{dy}{dx} = y(1 - x)$

D $\dfrac{dy}{dx} = y(x - 1)$

Question 8

The volume of a regular tetrahedron (equilateral triangular pyramid) is given by

$$V = \dfrac{s^3}{6\sqrt{2}},$$

where s is the length of each edge.

At what rate is the volume increasing when the length of each edge is 18 cm and all sides are increasing at a rate of 2 cm/s?

A $162\sqrt{2}\ \text{cm}^3/\text{s}$

B $324\sqrt{2}\ \text{cm}^3/\text{s}$

C $486\sqrt{2}\ \text{cm}^3/\text{s}$

D $972\sqrt{2}\ \text{cm}^3/\text{s}$

Question 9

Eight teams out of 16 are chosen to host a game of rugby league. Two of these teams are chosen to play on Friday, three on Saturday, and the rest on Sunday.

In how many ways can this be done?

A $\dfrac{16!}{6!3!3!2!}$

B $\dfrac{16!}{8!3!3!2!}$

C $\dfrac{16!}{8!6!3!2!}$

D $\dfrac{16!}{8!6!3!3!}$

Question 10

In which diagram is $\left|\text{proj}_{\underset{\sim}{b}}\underset{\sim}{a}\right| > \left|\underset{\sim}{b}\right|$?

A

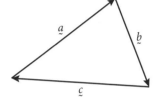

B

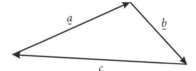

C

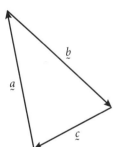

D

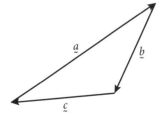

Section II

60 marks
Attempt Questions 11–14
Allow about 1 hours and 45 minutes for this section

- Answer the questions in the spaces provided. These spaces provide guidance for the expected length of response.
- Your responses should include relevant mathematical reasoning and/or calculations.

Question 11 (15 marks)

a   Evaluate $\int_{-1}^{1} \dfrac{dx}{\sqrt{2 - x^2}}$.   2 marks

b   Solve $|2x + 3| \geq 5$.   2 marks

c   Differentiate $\cos^{-1}\sqrt{1 - x^2}$.   3 marks

Question 11 continues on page 127

Question 11 (continued)

d Solve $\dfrac{dy}{dx} = 2x\cos^2 y$, giving y in terms of x.   2 marks

e What is the repeated root of $P(x) = x^3 - 8x^2 - 35x + 294$?   2 marks

f By expressing $5\sin\theta + 12\cos\theta$ in the form $R\sin(\theta + \alpha)$, solve $5\sin\theta + 12\cos\theta = 8$ for $0 \le \theta \le 2\pi$,   4 marks
correct to two decimal places.

End of Question 11

Question 12 (15 marks)

a   ©NESA 2002 HSC EXAM, QUESTION 4(a)

Lyndal hits the target on average 2 out of every 3 shots in archery competitions.

During a competition she has 10 shots at the target.

i   What is the probability that Lyndal hits the target exactly 9 times?   1 mark
Leave your answer in unsimplified form.

ii   What is the probability that Lyndal hits the target fewer than 9 times?   2 marks
Leave your answer in unsimplified form.

b   ©NESA 2000 HSC EXAM, QUESTION 2(c)

Solve the equation $\cos 2\theta = \sin \theta$, $0 \le \theta \le 2\pi$.   4 marks

Question 12 continues on page 129

Question 12 (continued)

c   A cylindrical water tank, with cross-sectional area 1 m² and height 1 m, has a 10 cm² hole   3 marks
at the bottom.

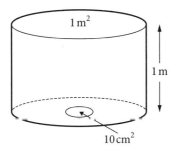

Torricelli's law says the volume of water in the tank is modelled by the differential equation

$$\frac{dV}{dt} = -a\sqrt{2gh},$$

where V is the volume of water in m³, t is time in seconds, a is the area of the hole in m², g is the
gravitational acceleration in m/s² and h is the height of the water in metres. Assume that $g = 10$ m/s².

If the tank is initially full, how long (in minutes and seconds, to the nearest second) will it take
for it to empty?

Question 12 continues on page 130

Question 12 (continued)

d   Poppy arrives home and finds the temperature of her lounge room is 33°C. She turns on the air conditioner and sets it for 23°C. After 10 minutes, the temperature decreases to 28°C.

At time t (minutes), the temperature T (°C) decreases according to the equation

$$\frac{dT}{dt} = -k(T - 23),$$

where k is a positive constant.

i   Show that $T = 23 + Ae^{-kt}$ is a solution to this equation, where A is a positive constant.   1 mark

ii   Find the values of A and k. Express k correct to four significant figures.   2 marks

iii   How long, to the nearest minute, will it take for the temperature to decrease to 24°C?   2 marks

End of Question 12

Question 13 (15 marks)

a   Use the substitution $u = 1 - x$ to evaluate $\int_0^1 x^2 \sqrt{1 - x}\, dx$ as an exact value.   3 marks

b   The shaded region shown is bounded by the graphs of $y = 2\sin^{-1} x$ and $y = \pi\sqrt{x}$.   4 marks

Find in exact form the volume of the solid of revolution formed when the region is rotated around the y-axis.

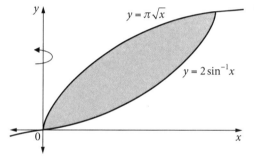

$y = \pi\sqrt{x}$

$y = 2\sin^{-1}x$

Question 13 continues on page 132

Question 13 (continued)

c   A rugby league ball is kicked from the ground with a velocity of 30 m/s from a point 20 m from the goal. The player wishes to kick the ball as far above the crossbar of the goal as possible, which is 3 m above ground level.

Let α be the angle of inclination that the ball is kicked and assume the acceleration due to gravity is $g = 10$ m/s^2.

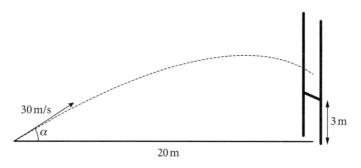

NOT TO SCALE

30 m/s

α

3 m

20 m

The displacement vector for the ball is $\underset{\sim}{r} = \begin{pmatrix} 30t\cos\alpha \\ 30t\sin\alpha - 5t^2 \end{pmatrix}$. Do NOT prove this.

i   Show that the ball will pass $\left(20\tan\alpha - \dfrac{20}{9}\tan^2\alpha - \dfrac{47}{9}\right)$ metres above the crossbar.   2 marks

ii   Hence, find to the nearest degree the angle that will give the maximum height above the crossbar.   3 marks

Question 13 continues on page 133

Question 13 (continued)

d Shella and Nordin play a game where they each flip 20 coins and count the number of heads.

Let X and Y represent the number of heads Shella and Nordin flip respectively.

    i Show that $\mathrm{Var}(X) = 5$.   1 mark

    ii Shella's score will be X^2, whereas Nordin's score will be $10Y$.   2 marks

       Whose score has the higher expected value?

End of Question 13

Question 14 (15 marks)

a   **i**   Show that $3x^3 + 14x^2 + 19x + 8 = (x + 1)^2(3x + 8)$.   1 mark

ii   Hence, prove by mathematical induction that for all integers $n \geq 1$,   4 marks

$$\frac{1}{3} + \frac{1}{8} + \frac{1}{15} + \cdots + \frac{1}{n(n + 2)} = \frac{n(3n + 5)}{4(n + 1)(n + 2)}.$$

Question 14 continues on page 135

Question 14 (continued)

b Let *OPQR* be a kite defined by the vectors $\underset{\sim}{p} = \overrightarrow{OP}$ and $\underset{\sim}{r} = \overrightarrow{OR}$.

Thus, $\left|\overrightarrow{QP}\right| = \left|\overrightarrow{OP}\right|$ and $\left|\overrightarrow{QR}\right| = \left|\overrightarrow{OR}\right|$. Let $\underset{\sim}{q} = \overrightarrow{OQ}$.

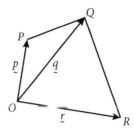

i Use the fact that for any vector $\underset{\sim}{v}, \left|\underset{\sim}{v}\right|^2 = \underset{\sim}{v} \cdot \underset{\sim}{v}$, to prove that $\underset{\sim}{q} \cdot \underset{\sim}{q} = 2(\underset{\sim}{p} \cdot \underset{\sim}{q}) = 2(\underset{\sim}{r} \cdot \underset{\sim}{q})$.   3 marks

ii Hence, prove that *OQ* is perpendicular to *PR*.   2 marks

Question 14 continues on page 136

Question 14 (continued)

c ©NESA 2019 HSC EXAM, QUESTION 14(c)

The diagram shows the two curves $y = \sin x$ and $y = \sin(x - \alpha) + k$, where $0 < \alpha < \pi$ and $k > 0$.

The two curves have a common tangent at x_0 where $0 < x_0 < \dfrac{\pi}{2}$.

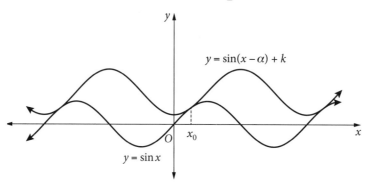

i Explain why $\cos x_0 = \cos(x_0 - \alpha)$.   1 mark

ii Show that $\sin x_0 = -\sin(x_0 - \alpha)$.   2 marks

Question 14 continues on page 137

Question 14 (continued)

   iii   Hence, or otherwise, find k in terms of α.   2 marks

END OF PAPER

WORKED SOLUTIONS

Section I (1 mark each)

Question 1

B The domain for ln is all positive numbers, so
$$\sqrt{x} > 0$$
$$\therefore x > 0.$$

This is a Year 11 understanding question.

Question 2

C Sum of roots $= -\dfrac{-3}{1} = 3$

This question is from the Year 11 Polynomials topic.

Question 3

D $\displaystyle\int \cos^2 4x \, dx = \int \frac{1}{2}(1 + \cos 8x) \, dx$

$\qquad\qquad = \dfrac{1}{2}x + \dfrac{1}{16}\sin 8x + c$

Use the HSC exam reference sheet to convert the square of a cosine for integration using a trigonometric identity.

Question 4

B $\begin{pmatrix} 2 \\ k \end{pmatrix} \cdot \begin{pmatrix} -1 \\ k-1 \end{pmatrix} = -2 + k(k-1) = 0$

$\qquad\qquad -2 + k^2 - k = 0$

$\qquad\qquad k^2 - k - 2 = 0$

$\qquad\qquad (k-2)(k+1) = 0$

$\qquad\qquad\qquad\qquad k = -1, 2$

This is a common vectors question involving the scalar product.

Question 5

D Test points, such as $x = 1$ or $x = -1$:

$$y(1) = \frac{\pi}{2} - \cos^{-1}(-1)$$

$$= \frac{\pi}{2} - \pi$$

$$= -\frac{\pi}{2}$$

If you know the graph of $y = \cos^{-1} x$, you can also reflect it in the y-axis, reflect it in the x-axis, then translate it up $\dfrac{\pi}{2}$ units.

Question 6

B $\qquad\qquad 4\sin^2 x = 4\sin x + 3$

$\qquad\qquad 4\sin^2 x - 4\sin x - 3 = 0$

$\qquad\qquad (2\sin x + 1)(2\sin x - 3) = 0$

$\sin x = \dfrac{3}{2}$ (no solution) or $\sin x = -\dfrac{1}{2}$

$$x = \frac{7\pi}{6}, \frac{11\pi}{6}, \frac{19\pi}{6}, \ldots$$

Make sure you choose the correct quadrants for the solutions.

Question 7

A At $(1, 0)$, $\dfrac{dy}{dx} > 0$, so option A.

In 4th quadrant, all gradients positive,

$\dfrac{dy}{dx} > 0$ for $x > 0$, $y < 0$, so option A.

Look for features to distinguish between choices.

Question 8

A $V = \dfrac{s^3}{6\sqrt{2}}$

Need to find $\dfrac{dV}{dt}$ when $s = 18$ and $\dfrac{ds}{dt} = 2$.

$$\frac{dV}{dt} = \frac{dV}{ds} \times \frac{ds}{dt}$$

$$\frac{dV}{ds} = \frac{s^2}{2\sqrt{2}}$$

$$\frac{dV}{dt} = \frac{s^2}{2\sqrt{2}} \times 2$$

$$= \frac{18^2}{2\sqrt{2}} \times 2$$

$$= \frac{324}{\sqrt{2}}$$

$$= \frac{324\sqrt{2}}{2}$$

$$= 162\sqrt{2} \text{ cm}^3/\text{s}$$

This is a Year 11 question involving related rates and the chain rule.

Question 9

B $\dbinom{16}{8} \times \dbinom{8}{2} \times \dbinom{6}{3} \times \dbinom{3}{3} = \dfrac{16!}{8!8!} \times \dfrac{8!}{2!6!} \times \dfrac{6!}{3!3!} \times \dfrac{3!}{3!0!}$

$\qquad\qquad\qquad\qquad\quad = \dfrac{16!}{8!3!3!2!}$

This is a harder Year 11 combinatorics question – use the HSC exam reference sheet.

Question 10

D This question is asking for when the length of the projection of $\underset{\sim}{a}$ on $\underset{\sim}{b}$ is greater than the length of $\underset{\sim}{b}$. The answer is D because if a perpendicular is drawn from $\underset{\sim}{a}$ to $\underset{\sim}{b}$, it will extend further than $\underset{\sim}{b}$.

This is a harder question that needs complete visual understanding of vector projection. It can help to rotate the paper to make $\underset{\sim}{b}$ the base to draw/visualise the projections.

Section II ($\checkmark$ = 1 mark)

Question 11 (15 marks)

a $\displaystyle\int_{-1}^{1} \dfrac{dx}{\sqrt{2-x^2}} = \left[\sin^{-1}\dfrac{x}{\sqrt{2}}\right]_{-1}^{1}$ $\checkmark$

$\qquad\qquad\qquad = \dfrac{\pi}{4} - \left(-\dfrac{\pi}{4}\right)$

$\qquad\qquad\qquad = \dfrac{\pi}{2}$ $\checkmark$

This is a standard integral involving an inverse trigonometric function. Use the HSC exam reference sheet.

b $|2x+3| \geq 5$

$2x+3 \geq 5$ or $2x+3 \leq -5$ $\checkmark$

So $x \geq 1$ or $x \leq -4$. $\checkmark$

This is a common type of Year 11 algebra question.

c $\dfrac{d}{dx}\sqrt{1-x^2} = -\dfrac{x}{\sqrt{1-x^2}}$ $\checkmark$

$\therefore \dfrac{d}{dx}\cos^{-1}\left(\sqrt{1-x^2}\right)$

$= -\dfrac{x}{\sqrt{1-x^2}} \times \left(-\dfrac{1}{\sqrt{1-(1-x^2)}}\right)$ $\checkmark$

$= \dfrac{x}{\sqrt{1-x^2}} \times \dfrac{1}{x}$

$= \dfrac{1}{\sqrt{1-x^2}}$ $\checkmark$

This is a complicated differentiation involving an inverse trigonometric function and the chain rule. Use the HSC exam reference sheet.

d $\qquad \dfrac{dy}{dx} = 2x\cos^2 y$

$\int \sec^2 y\, dy = \int 2x\, dx$

$\qquad \tan y = x^2 + c$ $\checkmark$

$\qquad\ \text{So } y = \tan^{-1}(x^2 + c)$ $\checkmark$

This is a common differential equation question involving separation of variables.

e $P(x) = x^3 - 8x^2 - 35x + 294$

$P'(x) = 3x^2 - 16x - 35$

Solve $P'(x) = 0$:

$3x^2 - 16x - 35 = 0$

$(3x+5)(x-7) = 0$

$\qquad\qquad\qquad x = -\dfrac{5}{3}$ or 7 $\checkmark$

$P(7) = 7^3 - 8(7^2) - 35(7) + 294$

$\qquad = 0$

Hence, the repeated root is $x = 7$. $\checkmark$

This is a Year 11 polynomial question involving the multiple root theorem and the factor theorem. Make sure to check the roots in the original polynomial $P(x)$, not in $P'(x)$.

f   Let $5\sin\theta + 12\cos\theta = R\sin(\theta + \alpha)$
$$= R(\sin\theta\cos\alpha + \cos\theta\sin\alpha)$$
$$= R\sin\theta\cos\alpha + R\cos\theta\sin\alpha$$

$\therefore R\cos\alpha = 5$ [1]

$\therefore R\sin\alpha = 12$ [2]

[2] $\div$ [1]: $\tan\alpha = \dfrac{12}{5}$

$\alpha = \tan^{-1}\dfrac{12}{5}$ ✓

$[1]^2 + [2]^2$: $R^2\cos^2\alpha + R^2\sin^2\alpha = 5^2 + 12^2$
$$R^2(\cos^2\alpha + \sin^2\alpha) = 169$$
$$R = 13$$

$\therefore 5\sin\theta + 12\cos\theta = 13\sin\left(\theta + \tan^{-1}\dfrac{12}{5}\right) = 8$ ✓

$\sin\left(\theta + \tan^{-1}\dfrac{12}{5}\right) = \dfrac{8}{13}$

$\theta + 1.176\,00\ldots = 0.662\,87\ldots$ ✓

$\theta = -0.513\,13\ldots$

But $0 \le \theta \le 2\pi$

so $1.176\,00\ldots \le \theta + 1.176\,00\ldots \le 2\pi + 1.176\,00\ldots$

$1.176\,00\ldots \le \theta + 1.176\,00\ldots \le 7.459\,18\ldots$

$\therefore \theta + 1.176\,00\ldots = \pi - 0.662\,87\ldots$ or $2\pi + 0.662\,87\ldots$

(2nd quadrant)   (1st quadrant again)

$\theta + 1.176\,00\ldots = 2.478\,72\ldots$ or $6.946\,05\ldots$

$\approx 1.30, 5.77$ ✓

> Auxiliary angle problems are very common in the HSC exam (see topic grid on page 40), although usually scaffolded (guiding steps given). Be very careful with the domain.

Question 12

a   **i** $P(X=9) = \dbinom{10}{9}\left(\dfrac{2}{3}\right)^9\left(\dfrac{1}{3}\right)^1$ ✓

> This is a very common type of binomial probability question. 'Unsimplified form' means you don't need to evaluate the expression.

ii $P(X<9) = 1 - P(X=10) - P(X=9)$ ✓
$$= 1 - \left(\dfrac{2}{3}\right)^{10} - \dbinom{10}{9}\left(\dfrac{2}{3}\right)^9\left(\dfrac{1}{3}\right)^1 \ ✓$$

> This question was in the 2002 HSC exam. Always look for shortcuts in probability problems. 'Fewer than 9' means 'not 9 or 10'. Always use part **i** in part **ii**. One common student error was to read 'fewer than 9' as '9 or fewer' (not the same thing).

b
$$\cos 2\theta = \sin\theta$$
$$1 - 2\sin^2\theta = \sin\theta$$
$$2\sin^2\theta + \sin\theta - 1 = 0 \ ✓$$
$$(2\sin\theta - 1)(\sin\theta + 1) = 0 \ ✓$$

Hence, $\sin\theta = \dfrac{1}{2}$ or -1. ✓

So $\theta = \dfrac{\pi}{6}, \dfrac{5\pi}{6}, \dfrac{3\pi}{2}$. ✓

> This is a typical trigonometric equation. Always change into a single trigonometric ratio first, using trigonometric identities, and know your exact values.

c $a = 10\,\text{cm}^2 = (10 \div 10\,000)\,\text{m}^2 = 0.001\,\text{m}^2, g = 10$

$$\frac{dV}{dt} = -a\sqrt{2gh}$$
$$= -0.001\sqrt{20h}$$
$$= -0.002\sqrt{5h}$$

$V = Ah = (1)h = h$

$$\frac{dV}{dh} = 1 \checkmark$$

Hence, $\dfrac{dt}{dh} = \dfrac{dt}{dV} \times \dfrac{dV}{dh}$

$$= \frac{1}{\dfrac{dV}{dt}} \times 1$$

$$= -\frac{1}{0.002\sqrt{5h}}$$

$$= -\frac{\sqrt{5}}{0.01\sqrt{h}}$$

$$= -\frac{100\sqrt{5}}{\sqrt{h}}.$$

$$\therefore t = \int_1^0 -\frac{100\sqrt{5}}{\sqrt{h}}\,dh \checkmark$$

$$= \left[-200\sqrt{5h}\right]_1^0$$
$$= 0 - \left(-200\sqrt{5}\right)$$
$$= 200\sqrt{5}$$
$$\approx 447\,\text{s}$$
$$= 7\,\text{min}\,27\,\text{s}$$

Hence, it will take approximately 7 min 27 s for the tank to empty. ✓

This is a Year 11 chain rule question. Change the equations into just 2 variables (one of them being time) and then integrate from beginning to end. Ensure all measurements are consistent in their units, and notice that we can technically also write $V = A$ (with $h = 1$ as the initial condition) but this wouldn't help with obtaining $\dfrac{dt}{dh}$.

d i
$$T = 23 + Ae^{-kt}$$
$$\frac{dT}{dt} = -kAe^{-kt}$$
$$-k(T - 23) = -k(23 + Ae^{-kt} - 23)$$
$$= -k(Ae^{-kt})$$
$$= \frac{dT}{dt} \checkmark$$

Newton's law of cooling is a common type of question, so learn this proof precisely. This question is also doable by integrating the differential equation.

ii When $t = 0$, $T = 33$:

$$33 = 23 + Ae^{-k(0)}$$
$$10 = A \checkmark$$

$$\therefore T = 23 + 10e^{-kt}$$

When $t = 10$, $T = 28$:

$$28 = 23 + 10e^{-k(10)}$$
$$5 = 10e^{-k(10)}$$
$$e^{-10k} = \frac{1}{2}$$
$$-10k = \ln\left(\frac{1}{2}\right)$$
$$k = -\frac{1}{10}\ln\left(\frac{1}{2}\right)$$
$$\approx 0.069\,31 \checkmark$$

A is the initial temperature, so substitute $t = 0$ to find its value. Then substitute the other condition to find k. Keep the value of k in your calculator for part **iii**.

iii When $t = 24$:

$$24 = 23 + 10e^{-kt}$$
$$1 = 10e^{-kt}$$
$$0.1 = e^{-kt} \checkmark$$
$$-kt = \ln 0.1$$
$$t = -\frac{1}{k}\ln 0.1$$
$$\approx 33.22$$

It will take about 33 minutes. ✓

Check that your answer seems reasonable. The question states that after 10 minutes, the temperature decreases to 28°C.

Question 13

a $u = 1 - x$

$du = -dx$

$dx = -du$ ✓

When $x = 0$, $u = 1$.

When $x = 1$, $u = 0$.

$\therefore \int_0^1 x^2 \sqrt{1-x}\, dx = \int_1^0 -(1-u)^2 \sqrt{u}\, du$

$= \int_0^1 (1 - 2u + u^2)\sqrt{u}\, du$

$= \int_0^1 \left(u^{\frac{1}{2}} - 2u^{\frac{3}{2}} + u^{\frac{5}{2}}\right) du$ ✓

$= \left[\frac{2}{3}u^{\frac{3}{2}} - \frac{4}{5}u^{\frac{5}{2}} + \frac{2}{7}u^{\frac{7}{2}}\right]_0^1$

$= \frac{2}{3} - \frac{4}{5} + \frac{2}{7} - 0$

$= \frac{16}{105}$ ✓

> This is integration by substitution. Be prepared to integrate powers that are fractions. When changing the limits of integration, maintain the order that they appeared in the original integral.

b Find points of intersection:

$\pi\sqrt{x} = 2\sin^{-1} x$

$(0, 0)$ and $(1, \pi)$ (by inspection and graph) ✓

Make x the subject of both functions:

$y = 2\sin^{-1} x$ $\qquad$ $y = \pi\sqrt{x}$

$x = \sin\left(\frac{y}{2}\right)$ $\qquad$ $x = \left(\frac{y}{\pi}\right)^2$ ✓

$V = \pi \int_0^\pi \left(x_1^2 - x_2^2\right) dy$

$= \pi \int_0^\pi \left(\sin^2\left(\frac{y}{2}\right) - \left(\frac{y}{\pi}\right)^4\right) dy$ ✓

$= \pi \int_0^\pi \left(\frac{1}{2}(1 - \cos y) - \frac{y^4}{\pi^4}\right) dy$

$= \pi \left[\frac{1}{2}y - \frac{1}{2}\sin y - \frac{y^5}{5\pi^4}\right]_0^\pi$

$= \pi \left[\frac{\pi}{2} - \frac{1}{2}\sin \pi - \frac{\pi^5}{5\pi^4} - 0\right]$

$= \pi \left[\frac{\pi}{2} - 0 - \frac{\pi}{5}\right]$

$= \frac{3\pi^2}{10}$ units^3 ✓

> A common error is using $(x_1 - x_2)^2$ instead of $x_1^2 - x_2^2$.

c **i** When $x = 20$,

$30t\cos \alpha = 20$

$t = \frac{2}{3}\sec \alpha$ ✓

Hence, $y = 30\left(\frac{2}{3}\sec \alpha\right)\sin \alpha - 5\left(\frac{2}{3}\sec \alpha\right)^2$

$= 20\tan \alpha - \frac{20}{9}\sec^2 \alpha$

Therefore, it will pass a distance in metres above the crossbar of:

$h = 20\tan \alpha - \frac{20}{9}(\tan^2 \alpha + 1) - 3$

$= 20\tan \alpha - \frac{20}{9}\tan^2 \alpha - \frac{47}{9}$ ✓

> Make sure you read the question properly and check that you have answered it. If you got the constant/last term incorrect, check that you haven't forgotten to subtract 3 due to the crossbar.

ii Find the angle α that maximises h.

$\frac{dh}{d\alpha} = 20\sec^2 \alpha - \frac{40}{9}\sec^2 \alpha \tan \alpha = 0$ ✓

$\sec \alpha = \frac{1}{\cos \alpha} \neq 0$, so dividing both sides by $\sec^2 \alpha$:

$20 - \frac{40}{9}\tan \alpha = 0$

$\tan \alpha = \frac{9}{2}$ ✓

$\alpha \approx 77.47°$

Hence, the angle that will give the maximum height is $77°$ to the nearest degree. ✓

> This is solving a projectile motion problem with a trigonometric equation.

d **i** $\text{Var}(X) = np(1 - p)$

$= 20 \times \frac{1}{2} \times \frac{1}{2}$

$= 5$ ✓

> This is a binomial distribution. Use the HSC exam reference sheet.

ii $E(10Y) = 10 \times 10 = 100$ ✓

$E(X^2) = \text{Var}(X) + \mu^2$ (based on the reference sheet)

$\qquad = 5 + 10^2$

$\qquad = 105$

Hence, Shella has a higher expected score. ✓

By using the HSC exam reference sheet, you can see that $E(X^2)$ can be calculated.

Question 14

a i RHS $= (x + 1)^2(3x + 8)$

$\qquad = (x^2 + 2x + 1)(3x + 8)$

$\qquad = 3x^3 + 6x^2 + 3x + 8x^2 + 16x + 8$

$\qquad = 3x^3 + 14x^2 + 19x + 8$

$\qquad = $ LHS ✓

It is often much easier to show that RHS = LHS, in this case, expanding rather than factorising.
This is a straightforward algebra question.

ii Let $P(n)$ be the statement $\dfrac{1}{3} + \dfrac{1}{8} + \dfrac{1}{15} + \cdots + \dfrac{1}{n(n + 2)} = \dfrac{n(3n + 5)}{4(n + 1)(n + 2)}$.

$P(1)$ is LHS $= \dfrac{1}{3}$, RHS $= \dfrac{1(3(1) + 5)}{4(1 + 1)(1 + 2)} = \dfrac{1}{3} = $ LHS

$\therefore P(1)$ is true. ✓

Assume $P(k)$ is true.

$P(k)$ is $\dfrac{1}{3} + \dfrac{1}{8} + \dfrac{1}{15} + \cdots + \dfrac{1}{k(k + 2)} = \dfrac{k(3k + 5)}{4(k + 1)(k + 2)}$.   [*]

RTP $P(k + 1)$: $\dfrac{1}{3} + \dfrac{1}{8} + \dfrac{1}{15} + \cdots + \dfrac{1}{k(k + 2)} + \dfrac{1}{(k + 1)(k + 3)} = \dfrac{(k + 1)(3(k + 1) + 5)}{4(k + 2)(k + 3)}$

$\qquad\qquad\qquad\qquad\qquad\qquad\qquad\qquad\qquad\qquad = \dfrac{(k + 1)(3k + 8)}{4(k + 2)(k + 3)}$

LHS $= \dfrac{k(3k + 5)}{4(k + 1)(k + 2)} + \dfrac{1}{(k + 1)(k + 3)}$ from [*] ✓

$\quad = \dfrac{k(3k + 5)(k + 3)}{4(k + 1)(k + 2)(k + 3)} + \dfrac{4(k + 2)}{4(k + 1)(k + 2)(k + 3)}$

$\quad = \dfrac{3k^3 + 14k^2 + 15k + 4k + 8}{4(k + 1)(k + 2)(k + 3)}$

$\quad = \dfrac{3k^3 + 14k^2 + 19k + 8}{4(k + 1)(k + 2)(k + 3)}$ ✓

$\quad = \dfrac{(k + 1)^2(3k + 8)}{4(k + 1)(k + 2)(k + 3)}$ (from part **i**)

$\quad = \dfrac{(k + 1)(3k + 8)}{4(k + 2)(k + 3)}$

$\quad = $ RHS

$P(k + 1)$ is true.

So by mathematical induction, $P(n)$ is true for all integers $n \geq 1$. ✓

Be very careful with expanding, simplifying and adding fractions. Always use part **i** in part **ii**. In the inductive step, working out the RHS for $P(k + 1)$ first helps in figuring out where to take the direction of your working from the LHS.

b i $\overrightarrow{QP} = \underset{\sim}{p} - \underset{\sim}{q}$

Hence, $\left|\underset{\sim}{p}\right| = \left|\overrightarrow{QP}\right| = \left|\underset{\sim}{p} - \underset{\sim}{q}\right|$.

Therefore, $\underset{\sim}{p} \cdot \underset{\sim}{p} = \left|\underset{\sim}{p}\right|^2$

$\qquad = \left|\underset{\sim}{p} - \underset{\sim}{q}\right|^2$

$\qquad = (\underset{\sim}{p} - \underset{\sim}{q}) \cdot (\underset{\sim}{p} - \underset{\sim}{q})$

$\qquad = \underset{\sim}{p} \cdot \underset{\sim}{p} - 2\underset{\sim}{p} \cdot \underset{\sim}{q} + \underset{\sim}{q} \cdot \underset{\sim}{q}.$ ✔

Thus, $\underset{\sim}{q} \cdot \underset{\sim}{q} = 2(\underset{\sim}{p} \cdot \underset{\sim}{q})$. ✔

$\overrightarrow{QR} = \underset{\sim}{r} - \underset{\sim}{q}$

Hence, $\left|\underset{\sim}{r}\right| = \left|\overrightarrow{QR}\right| = \left|\underset{\sim}{r} - \underset{\sim}{q}\right|$.

Therefore, by the same argument,
$\underset{\sim}{q} \cdot \underset{\sim}{q} = 2(\underset{\sim}{r} \cdot \underset{\sim}{q}) = 2(\underset{\sim}{p} \cdot \underset{\sim}{q})$. ✔

> This is a harder question that requires the ability to properly manipulate vector expressions.

ii $\overrightarrow{OQ} \cdot \overrightarrow{PR} = \underset{\sim}{q} \cdot (\underset{\sim}{r} - \underset{\sim}{p})$

$\qquad = \underset{\sim}{r} \cdot \underset{\sim}{q} - \underset{\sim}{p} \cdot \underset{\sim}{q}$ ✔

But from part **i**: $\underset{\sim}{p} \cdot \underset{\sim}{q} = \underset{\sim}{r} \cdot \underset{\sim}{q}$

Hence, $\overrightarrow{OQ} \cdot \overrightarrow{PR} = 0$ and so OQ is perpendicular to PR. ✔

> This follows from part **i** and the vector definition of perpendicular.

c i Gradient of tangent at $y = \sin(x - \alpha) + k$ at $x = x_0$:

$\dfrac{dy}{dx} = \cos(x - \alpha) = \cos(x_0 - \alpha)$

Gradient of tangent at $y = \sin x$ at $x = x_0$:

$\dfrac{dy}{dx} = \cos x = \cos x_0$

Common tangent, so the gradients are equal.

$\cos(x_0 - \alpha) = \cos x_0$ ✔

> This was also the final question for the 2019 HSC exam, so it was quite difficult and unusual as it required both deep and precise knowledge of calculus with trigonometric equations. Students did not do well because they ran out of time or because they were struggling to understand the question.

ii Since $\cos(x_0 - \alpha) = \cos x_0$ and x_0 is in the 1st quadrant $(0 < x_0 < \frac{\pi}{2})$:

$\cos x_0 > 0$

and $x_0 - \alpha$ must be in the 4th quadrant (where cos is also positive)

$\therefore x_0 - \alpha = -x_0 \quad [*]$ ✔

$\therefore \sin(x_0 - \alpha) = \sin(-x_0)$

$\qquad = -\sin x_0$

$\sin x_0 = -\sin(x_0 - \alpha)$, as required. ✔

> This atypical proof required the link to part **i** and some logical reasoning using the properties of the cosine and sine functions, particularly their symmetry (even and odd functions respectively) in the 1st and 4th quadrants. You should note that there are different methods of solving this problem, not just the one provided above.

iii As both graphs touch at $x = x_0$:

$\sin(x_0 - \alpha) + k = \sin x_0$

Substituting equation from part **ii** into $\sin(x_0 - \alpha)$:

$-\sin x_0 + k = \sin x_0$

$\qquad k = 2\sin x_0$ ✔

But from $[*]$:

$x_0 - \alpha = -x_0$

$2x_0 = \alpha$

$x_0 = \dfrac{\alpha}{2}$

$\therefore k = 2\sin\left(\dfrac{\alpha}{2}\right)$ ✔

> This was the hardest part of the question and it should not be surprising that you need to use your answers to parts **i** and **ii** to find the answer. There was a heavy reliance of algebraic proofs involving trigonometric properties.

The 2020 Mathematics Extension 1 HSC Exam Worked Solutions

The 2020 HSC exam and other past HSC papers can be downloaded from the NESA website (www.educationstandards.nsw.edu.au) by selecting 'Year 11 – Year 12', 'HSC exam papers'. NESA marking feedback and guidelines can also be found there.

Section I (1 mark each)

Question 1

A $x^2 - 2x - 3 \geq 0$

$(x - 3)(x + 1) \geq 0$

$x \leq -1$ or $x \geq 3$

This is a straightforward Year 11 question.

Question 2

C   Domain and range of $y = \sqrt{x}$ and $x \geq 0$
and $y \geq 0$.

Thus, domain and range of $y = 1 + \sqrt{x}$ is $x \geq 0$
and $y \geq 1$.

Hence, the domain and range of $f^{-1}(x)$ is
$y \geq 0$ and $x \geq 1$.

For the inverse function, swap the domain
and range.

Question 3

D $\dfrac{1}{4x^2 + 1} = \dfrac{1}{2} \times \dfrac{2}{1 + (2x)^2}$

Hence, the anti-derivative of $\dfrac{1}{4x^2 + 1}$ is
$\dfrac{1}{2} \tan^{-1}(2x) + c$.

Use the HSC exam reference sheet.

Question 4

B $(2\underset{\sim}{i} + 3\underset{\sim}{j}) + (3\underset{\sim}{i} - 2\underset{\sim}{j}) + (4\underset{\sim}{i} - 3\underset{\sim}{j}) = 9\underset{\sim}{i} - 2\underset{\sim}{j}$

$\left|9\underset{\sim}{i} - 2\underset{\sim}{j}\right| = \sqrt{9^2 + 2^2} = \sqrt{85}$

Drawing a sketch would be useful.

Question 5

C   Monic polynomials are concave up as $x \to \infty$
(options C and D).

$x^2 + x + 1$ has no real zeros so $p(x)$ only has
one zero (options A and C).

Eliminate options by using the clues in the
question.

Question 6

D   A: $\underset{\sim}{v} = \underset{\sim}{b} - \underset{\sim}{a}$ C: $\underset{\sim}{v} = -(\underset{\sim}{a} + \underset{\sim}{b})$

B: $\underset{\sim}{v} = \underset{\sim}{a} + \underset{\sim}{b}$ D: $\underset{\sim}{v} = \underset{\sim}{a} + (-\underset{\sim}{b})$

This is a visual understanding question.

Question 7

A   When $x = 0$, $\dfrac{dy}{dx} = 0$ ($y \neq 0$), so the dashes
should be flat on the y-axis (options A and B).

In the 2nd quadrant, x is negative, y is positive,
so $\dfrac{dy}{dx} > 0$ and all dashes should be increasing
(option A only).

Identify special features in the graphs.

Question 8

C $^{10}C_6 \, {}^6P_4 = \dfrac{10!}{6!4!} \times \dfrac{6!}{2!}$

$= \dfrac{10!}{4!2!}$

It is vital that you can interpret what the question
is asking.

Question 9

B $\begin{pmatrix} 6 \\ 7 \end{pmatrix} - \begin{pmatrix} 4 \\ 8 \end{pmatrix} = \begin{pmatrix} 4 \\ 8 \end{pmatrix} - \begin{pmatrix} x \\ y \end{pmatrix}$

$\begin{pmatrix} x \\ y \end{pmatrix} = \begin{pmatrix} 2 \\ 9 \end{pmatrix}$

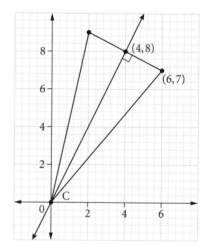

This is a difficult question conceptually. It is vital
that you draw a diagram to visualise this question.

Question 10

A $\dfrac{dR}{dt} < 0 \to \dfrac{dP}{dt} > 0 \to \dfrac{dQ}{dt} > 0$

Once you recognise that any number squared is
positive, this question is straightforward.

Section II ($\checkmark$ = 1 mark)

Question 11 (15 marks)

a i $P(2) = 2^3 + 3 \times 2^2 - 13 \times 2 + 6 = 0$ $\checkmark$

ii Since 2 is a zero, $A(x) = x - 2$. $\checkmark$

Hence, $B(x) = x^2 + 5x - 3$ by long division.

$$
\begin{array}{r}
x^2 + 5x - 3 \\
x - 2 \overline{\smash{)}\, x^3 + 3x^2 - 13x + 6} \\
\underline{x^3 - 2x^2} \\
5x^2 - 13x \\
\underline{5x^2 - 10x} \\
-3x + 6 \\
\underline{-3x + 6} \\
0
\end{array}
$$

So $P(x) = (x - 2)(x^2 + 5x - 3)$. $\checkmark$

This is a Year 11 polynomials question requiring the factor theorem. Note the link to part **i**. A common error was saying $A(x) = x + 2$, not $x - 2$.

b $\begin{pmatrix} a \\ -1 \end{pmatrix} \cdot \begin{pmatrix} 2a - 3 \\ 2 \end{pmatrix} = 0$ $\checkmark$

Hence, $a(2a - 3) + (-1)(2) = 0$

$2a^2 - 3a - 2 = 0$ $\checkmark$

$(2a + 1)(a - 2) = 0$.

So $a = -\dfrac{1}{2}$ or 2. $\checkmark$

Remember the scalar product property of perpendicular vectors. Some students were let down by incorrect use of the scalar product and incorrect solving of a simple quadratic equation.

c

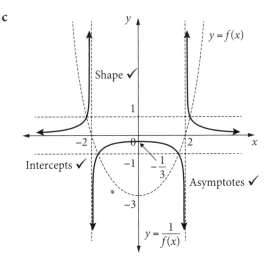

This is a typical Year 11 graphing question. It was generally well done. When graphing reciprocal functions, make sure that you include asymptotes and intercepts. Some students tried to find the equation of $f(x)$, which was not required. This question did not require any algebra. Some students also misinterpreted $\dfrac{1}{f(x)}$ as being the inverse function $f^{-1}(x)$: it isn't!

d Let $\sin x + 3 \cos x = A \sin (x + \alpha)$

$= A(\sin x \cos \alpha + \cos x \sin \alpha)$

$= A \sin x \cos \alpha + A \cos x \sin \alpha$

$\therefore A \cos \alpha = \sqrt{3}$ [1]

$\therefore A \sin \alpha = 3$ [2]

[2] ÷ [1]: $\tan \alpha = \dfrac{3}{\sqrt{3}} = \sqrt{3}$

$\alpha = \dfrac{\pi}{3}$ $\checkmark$

$[1]^2 + [2]^2$: $A^2 \cos^2 \alpha + A^2 \sin^2 \alpha = (\sqrt{3})^2 + 3^2$

$A^2(\cos^2 \alpha + \sin^2 \alpha) = 12$

$A = \sqrt{12} = 2\sqrt{3}$ $\checkmark$

$\therefore \sqrt{3} \sin x + 3 \cos x = 2\sqrt{3} \sin\left(x + \dfrac{\pi}{3}\right) = \sqrt{3}$

$\sin\left(x + \dfrac{\pi}{3}\right) = \dfrac{1}{2}$

Thus, $x + \dfrac{\pi}{3} = \dfrac{5\pi}{6}, \dfrac{13\pi}{6}$

$x = \dfrac{\pi}{2}$, $\checkmark$ $\dfrac{11\pi}{6}$ $\checkmark$

This is a typical 4-mark auxiliary angle question. Know your exact trigonometric ratios in radian form, and make sure you include all correct solutions for the given domain (such as $\dfrac{11\pi}{6}$).

e $\dfrac{dy}{dx} = e^{2y}$

$\dfrac{dx}{dy} = \dfrac{1}{e^{2y}}$ $\checkmark$

$= e^{-2y}$

Thus, $x = -\dfrac{1}{2}e^{-2y} + c.$ $\checkmark$

This is a simple differential equation. Note that this 2-mark question asked for x in terms of y, but some students wasted time going further to find y in terms of x.

Question 12 (14 marks)

a Let $P(n)$ be the statement $1 \times 2 + 2 \times 5 + \cdots + n(3n - 1) = n^2(n + 1)$.

$P(1)$ is $1 \times 2 = 1^2(1 + 1)$, which is true (both sides = 2). ✓

Assume $P(k)$ is true.

$P(k)$ is $1 \times 2 + 2 \times 5 + \cdots + k(3k - 1) = k^2(k + 1)$ [*]

$P(k + 1)$ is $1 \times 2 + 2 \times 5 + \cdots + k(3k - 1) + (k + 1)(3[k + 1] - 1) = (k + 1)^2(k + 1 + 1)$

RTP: $1 \times 2 + 2 \times 5 + \cdots + k(3k - 1) + (k + 1)(3k + 2) = (k + 1)^2(k + 2)$

$$\begin{aligned} \text{LHS} &= k^2(k + 1) + (k + 1)(3k + 2) \quad \text{using [*]} \checkmark \\ &= (k + 1)(k^2 + 3k + 2) \\ &= (k + 1)(k + 1)(k + 2) \\ &= (k + 1)^2(k + 2) \\ &= \text{RHS} \end{aligned}$$

Hence, $P(k + 1)$ is true and therefore by induction $P(n)$ is true for all integers $n \geq 1$. ✓

This is a straightforward induction question. It would not normally be so early in the exam paper. When proving $P(k + 1)$, better students knew to take out a factor of $(k + 1)$ rather than expand. You should also aim to start with the LHS (or RHS) and aim to prove that it is equal to the RHS (or LHS), not *start* with LHS = RHS and then manipulate *both sides* to make them equal (the HSC markers reported that many students did this in 2020).

b i $E(X) = np$

$$= 10 \times \frac{3}{5}$$

$$= 60 \checkmark$$

ii $\text{Var}(X) = np(1 - p)$

$$= 100 \times \frac{3}{5} \times \frac{2}{5}$$

$$= 24$$

Therefore, the standard deviation is $\sqrt{24} \approx 4.9$, which is approximately 5. ✓

iii From the reference sheet, the probability of being within 1 standard deviation of the mean, 60; that is, between 55 and 65, is 68%. ✓

Use the HSC exam reference sheet for all of part **b**.

c The number of ways of selecting 3 topics from 8 is $^8C_3 = 56$. ✓

Since $400 \div 56 = 7.1$, by the pigeonhole principle and by rounding up to 8, at least 8 students passed the same 3 topics. For *every* combination of 3 topics to be passed by 7 students each, this would require $7 \times 56 = 392$ students. But because there are 400 students (greater than 392), at least one of the combinations had 8 or more students. ✓

This is a pigeonhole principle question from Year 11, but you need to identify the use of combinations and show deep understanding. Learn the logic behind the principle, and the reason we round *up* the answer. Also learn how to 'explain' an answer in words using logic, as many students struggled here. Maths isn't just calculations and formulas.

d $\displaystyle\int_0^{\frac{\pi}{2}} \cos 5x \sin 3x \, dx$

$\displaystyle = \frac{1}{2} \int_0^{\frac{\pi}{2}} (\sin 8x - \sin 2x) \, dx$ ✓ (from the reference sheet)

$\displaystyle = \frac{1}{2} \left[-\frac{1}{8} \cos 8x + \frac{1}{2} \cos 2x \right]_0^{\frac{\pi}{2}}$ ✓

$\displaystyle = \frac{1}{2} \left(-\frac{1}{8} - \frac{1}{2} + \frac{1}{8} - \frac{1}{2} \right)$

$\displaystyle = -\frac{1}{2}$ ✓

Use the HSC exam reference sheet and recognise the trigonometric identities, in this case, the product to sums formulas. Also be careful with the integration and negative signs, as there were many careless errors made by students here. The question is unusual for having a negative answer.

e $y \, dy = -x \, dx$

Integrating: $\displaystyle \frac{1}{2} y^2 = -\frac{1}{2} x^2 + c$ ✓

When $x = 1$, $y = 0$:

$\displaystyle 0 = -\frac{1}{2} + c$

$\displaystyle c = \frac{1}{2}$ ✓

Thus, $\displaystyle \frac{1}{2} y^2 = -\frac{1}{2} x^2 + \frac{1}{2}$

$y^2 = 1 - x^2$.

So $x^2 + y^2 = 1$. ✓

This is a standard differential equation involving separation of variables. Some students complicated things by putting constants on *both* sides when integrating. This is unnecessary and one constant will do. Note also that the answer is not a function, but the equation of the unit circle.

Question 13 (16 marks)

a i $\displaystyle \frac{d}{d\theta}(\sin^3 \theta) = 3\sin^2 \theta \cos \theta$ ✓

This was a straightforward 1-mark question involving the chain rule, but some students complicated things by using the product rule or trigonometric identities.

ii $x = \tan \theta$

$dx = \sec^2 \theta \, d\theta$

When $x = 0$, $\theta = 1$.

When $x = 1$, $\theta = \dfrac{\pi}{4}$. ✓

$\displaystyle \therefore \int_0^1 \frac{x^2}{(1 + x^2)^{\frac{5}{2}}} \, dx = \int_0^{\frac{\pi}{4}} \frac{\tan^2 \theta}{(1 + \tan^2 \theta)^{\frac{5}{2}}} \sec^2 \theta \, d\theta$

$\displaystyle = \int_0^{\frac{\pi}{4}} \frac{\tan^2 \theta}{(\sec^2 \theta)^{\frac{5}{2}}} \sec^2 \theta \, d\theta$

$\displaystyle = \int_0^{\frac{\pi}{4}} \frac{\tan^2 \theta}{\sec^3 \theta} \, d\theta$ ✓

$\displaystyle = \int_0^{\frac{\pi}{4}} \frac{\sin^2 \theta}{\cos^2 \theta} \cos^3 \theta \, d\theta$

$\displaystyle = \int_0^{\frac{\pi}{4}} \sin^2 \theta \cos \theta \, d\theta$ ✓

$\displaystyle = \frac{1}{3} \left[\sin^3 \theta \right]_0^{\frac{\pi}{4}}$ (from part **i**)

$\displaystyle = \frac{1}{3} \left(\left(\frac{1}{\sqrt{2}} \right)^3 - 0^3 \right)$

$\displaystyle = \frac{1}{6\sqrt{2}}$

$\displaystyle = \frac{\sqrt{2}}{12}$ ✓

This is a typical substitution question, using part **i**. Also, don't forget to change the limits of integration.

b First, find the point of intersection of the two curves.

$$\cos 2x = \sin x$$
$$1 - 2\sin^2 x = \sin x$$
$$2\sin^2 x + \sin x - 1 = 0$$
$$(2\sin x - 1)(\sin x + 1) = 0$$
$$\sin x = -1, \frac{1}{2}$$

$x = \dfrac{\pi}{6}$ as $\sin x > 0$ for the point of intersection needed. ✓

Hence, $\displaystyle V = \pi \int_0^{\frac{\pi}{6}} (\cos^2 2x - \sin^2 x) \, dx$

$\displaystyle = \frac{\pi}{2} \int_0^{\frac{\pi}{6}} \left[(1 + \cos 4x) - (1 - \cos 2x) \right] dx$

$\displaystyle = \frac{\pi}{2} \int_0^{\frac{\pi}{6}} (\cos 4x + \cos 2x) \, dx$ ✓

$\displaystyle = \frac{\pi}{2} \left[\frac{1}{4} \sin 4x + \frac{1}{2} \sin 2x \right]_0^{\frac{\pi}{6}}$ ✓

$\displaystyle = \frac{\pi}{2} \left(\frac{1}{4} \sin \frac{2\pi}{3} + \frac{1}{2} \sin \frac{\pi}{3} - 0 \right)$

$\displaystyle = \frac{\pi}{2} \left(\frac{1}{4} \times \frac{\sqrt{3}}{2} + \frac{1}{2} \times \frac{\sqrt{3}}{2} \right)$

$\displaystyle = \frac{3\pi\sqrt{3}}{16}$ units^3 ✓

This unscaffolded 4-mark question gave no clues, but it involved simultaneous equations, double-angle formulas and integration of trigonometric functions. One common error was integrating $(y_1 - y_2)^2$ rather than $(y_1{}^2 - y_2{}^2)$.

c i $f(x) = \tan(\cos^{-1} x)$

$$f'(x) = \sec^2\left(\cos^{-1} x\right) \times \left(-\frac{1}{\sqrt{1-x^2}}\right) \checkmark$$

$$= -\frac{1}{\cos^2\left(\cos^{-1} x\right)\sqrt{1-x^2}}$$

$$= -\frac{1}{x^2\sqrt{1-x^2}} \checkmark$$

$$g(x) = \frac{\sqrt{1-x^2}}{x}$$

Using the quotient rule:

$$u = \sqrt{1-x^2} \qquad \frac{du}{dx} = \frac{1}{2}\left(1-x^2\right)^{-\frac{1}{2}}(-2x) = \frac{-x}{\sqrt{1-x^2}}$$

$$v = x \qquad \frac{dv}{dx} = 1$$

$$g'(x) = \frac{\dfrac{-x^2}{\sqrt{1-x^2}} - \sqrt{1-x^2}}{x^2} \checkmark$$

$$= \frac{-x^2 - \left(1-x^2\right)}{x^2\sqrt{1-x^2}}$$

$$= -\frac{1}{x^2\sqrt{1-x^2}} \checkmark$$

$$= f'(x)$$

This 4-mark question needs strong differentiation and algebra skills and noting that $\sec x = \dfrac{1}{\cos x}$.

ii The domain and range of both f and g are $[-1,0) \cup (0,1]$ and $(-\infty, \infty)$ respectively. $\checkmark$

If $f'(x) = g'(x)$, then integrating both sides means that $f(x) = g(x) + c$; that is, they differ by a constant.

Substitute $x = 1$ into both functions:

$$f(1) = \tan(\cos^{-1} 1) = \tan 0 = 0$$

$$g(1) = \frac{\sqrt{1-1^2}}{1} = 0 = f(1)$$

But both functions are discontinuous at $x = 0$, so we need to be careful when integrating.

We should also test a point on the other side of the domain, such as $x = -1$:

$$f(-1) = \tan\left(\cos^{-1}(-1)\right) = \tan \pi = 0$$

$$g(-1) = \frac{\sqrt{1-(-1)^2}}{-1} = 0 = f(-1) \checkmark$$

(Note that $g(1)$ and $g(-1)$ can also be read off the graph to save time.)

So the constant is 0 and $f(x) = g(x)$. $\checkmark$

In this question, take care you show that the functions have the same domain and range and have the same value at both sides of the domain, in order to score the 3 marks. If you miss either of these, then you fail to show equality.

9780170459259

Question 14 (15 marks)

a   i $(1+x)^{2n} = \binom{2n}{0}1^{2n}x^0 + \binom{2n}{1}1^{2n-1}x^1 + \cdots + \binom{2n}{2n-1}1^1 x^{2n-1} + \binom{2n}{2n}1^0 x^{2n}$

$= \binom{2n}{0}x^0 + \binom{2n}{1}x^1 + \cdots + \binom{2n}{2n-1}x^{2n-1} + \binom{n}{n}x^{2n}$

$(1+x)^n = \binom{n}{0}x^0 + \binom{n}{1}x^1 + \cdots + \binom{n}{n-1}x^{n-1} + \binom{n}{n}x^n$

For $(1+x)^{2n}$, $\binom{2n}{n}$ is the coefficient of x^n. ✔

Need to find the coefficient of x^n in $(1+x)^n (1+x)^n$.

The terms that give x^n are the 1st term in the 1st $(1+x)^n$ times the last term in the 2nd $(1+x)^n$, the 2nd term in the 1st $(1+x)^n$ times the 2nd-last in the 2nd $(1+x)^n$, and so on:

$\binom{n}{0}\binom{n}{n}x^n + \binom{n}{1}\binom{n}{n-1}x^n + \cdots + \binom{n}{n-1}\binom{n}{1}x^n + \binom{n}{n}\binom{n}{0}x^n$

The coefficient of x^n in $(1+x)^n(1+x)^n$ is:

$\binom{n}{0}^n + \binom{n}{1}\binom{n}{n-1} + \cdots + \binom{n}{n}\binom{n}{0} = \binom{n}{0}\binom{n}{0} + \binom{n}{1}\binom{n}{1} + \cdots + \binom{n}{n}\binom{n}{n}$ (by symmetry)

$= \binom{n}{0}^2 + \binom{n}{1}^2 + \cdots + \binom{n}{n}^2$

Hence, by equating coefficients, $\binom{2n}{n} = \binom{n}{0}^2 + \binom{n}{1}^2 + \cdots + \binom{n}{n}^2$. ✔

Question 14 is quite challenging, which is not surprising because it's the final question in the exam. You should be able to equate coefficients between expressions. The HSC markers say to make sure that you show enough working and explanation to earn the 2 marks for a 'show that' question where the answer is given.

ii   Let $2k$ be the number of members chosen, with k being the number of women/men.

The number of ways of choosing k women and k men from n women and n men is:

$\binom{n}{k}\binom{n}{k} = \binom{n}{k}^2$ ✔

k ranges from 0 to n, so the total number of ways is $\binom{n}{0}^2 + \binom{n}{1}^2 + \cdots + \binom{n}{n}^2$,

which equals $\binom{2n}{n}$ from part **i**. ✔

This is another complex unfamiliar question in which looking for a general pattern was the key. It should be obvious that part **i** needs to be used in part **ii**. Giving reasons usually means using words as well as calculations. The markers reported many students writing mini-essays for a 2-mark question. Write out the pattern and then generalise it. Study worked solutions to past HSC questions for examples of explanations. Note that 0 is considered an odd number in this question.

WORK**E**D SOLUTIONS

iii If k is the number of women/men chosen, then note that $k \neq 0$. Hence, the number of ways of choosing the members and then choosing a woman and a man is:

$$\binom{n}{1}^2 \binom{1}{1}\binom{1}{1} + \binom{n}{2}^2 \binom{2}{1}\binom{2}{1} + \cdots + \binom{n}{n}^2 \binom{n}{1}\binom{n}{1} = \binom{n}{1}^2 \binom{1}{1}^2 + \binom{n}{2}^2 \binom{2}{1}^2 + \cdots + \binom{n}{n}^2 \binom{n}{1}^2 \checkmark$$

$$= \binom{n}{1}^2 1^2 + \binom{n}{2}^2 2^2 + \cdots + \binom{n}{n}^2 n^2 \checkmark$$

This is only straightforward if part **ii** could be completed. Otherwise, this question is harder than part **ii**.

iv The number of ways of choosing one woman and one man is $n \times n = n^2$.

This leaves $(n-1)$ women and $(n-1)$ men.

From part **ii**, the number of ways of choosing the rest is $\binom{2(n-1)}{n-1} = \binom{2n-2}{n-1}$. $\checkmark$

Hence, the total number of ways is $n^2 \binom{2n-2}{n-1}$.

Therefore, $1^2 \binom{n}{1}^2 + 2^2 \binom{n}{2}^2 + \cdots + n^2 \binom{n}{n}^2 = n^2 \binom{2n-2}{n-1}$. $\checkmark$

The instruction to use part **ii** is absolutely necessary in this question. Whether or not it could be proven is not as important as whether it was understood. This is quite a challenging binomial coefficients and combinatorics problem. Note how the parts **(i)–(iv)** of this question are related, and parts **(i)–(iii)** are 'show that' questions.

b i $\sin 3\theta = \sin 2\theta \cos \theta + \cos 2\theta \sin \theta$

$\qquad = 2 \sin \theta \cos^2 \theta + (1 - 2\sin^2 \theta) \sin \theta$

$\qquad = 2 \sin \theta (1 - \sin^2 \theta) + \sin \theta - 2 \sin^3 \theta$

$\qquad = 2 \sin \theta - 2 \sin^3 \theta + \sin \theta - 2 \sin^3 \theta$

$\qquad = 3 \sin \theta - 4 \sin^3 \theta \checkmark$

Thus, $4 \sin^3 \theta - 3 \sin \theta + \sin 3\theta = 0$.

Hence, $\sin^3 \theta - \dfrac{3}{4} \sin \theta + \dfrac{\sin 3\theta}{4} = 0$. $\checkmark$

With this question, take a deep breath and do some thinking and planning before writing. The HSC markers saw a lot of long, messy and confusing working, for a 2-mark question. Don't try to use all trigonometric identities at once, as some students did, which makes the working more complicated. Writing $\sin 3\theta$ in terms of powers of $\sin \theta$ or $\cos 3\theta$ in terms of powers of $\cos \theta$ is a common HSC exam question, so start with that.

ii Substituting, $64 \sin^3 \theta - 48 \sin \theta + 8 = 0$ $\checkmark$

But from part **i**, $64 \sin^3 \theta - 48 \sin \theta + 16 \sin 3\theta = 0$

Comparing both equations, $16 \sin 3\theta = 8$

$$\sin 3\theta = \frac{1}{2} \checkmark$$

Always look to use the previous parts of questions.
The exam writer is always giving you clues to be helpful.

WORKED SOLUTIONS

iii From part **ii**, $3\theta = \dfrac{\pi}{6}, \dfrac{5\pi}{6}, \dfrac{13\pi}{6}, \dfrac{17\pi}{6}, \dfrac{25\pi}{6}, \ldots$

Hence, $\theta = \dfrac{\pi}{18}, \dfrac{5\pi}{18}, \dfrac{13\pi}{18}, \dfrac{17\pi}{18}, \dfrac{25\pi}{18}, \ldots$

However, unique values of $\sin\theta$ are only given by $\theta = \dfrac{\pi}{18}, \dfrac{5\pi}{18}, \dfrac{25\pi}{18}$. ✓ (An important step!)

Thus, the roots of the cubic equation $x^3 + 0x^2 - 12x + 8 = 0$ in part **ii** are

$$x = 4\sin\dfrac{\pi}{18}, \; 4\sin\dfrac{5\pi}{18}, \; 4\sin\dfrac{25\pi}{18}.$$

From the reference sheet, $\alpha + \beta + \gamma = -\dfrac{b}{a} = -\dfrac{0}{1} = 0$ and $\alpha\beta + \alpha\gamma + \beta\gamma = \dfrac{c}{a} = \dfrac{-12}{1} = -12.$ ✓

Thus, $\alpha^2 + \beta^2 + \gamma^2 = (\alpha + \beta + \gamma)^2 - 2(\alpha\beta + \alpha\gamma + \beta\gamma)$
$$= 0^2 - 2 \times (-12)$$
$$= 24.$$

So $16\left(\sin^2\dfrac{\pi}{18} + \sin^2\dfrac{5\pi}{18} + \sin^2\dfrac{25\pi}{18}\right) = 24.$

Therefore, $\sin^2\dfrac{\pi}{18} + \sin^2\dfrac{5\pi}{18} + \sin^2\dfrac{25\pi}{18} = \dfrac{3}{2}.$ ✓

This is a very challenging question (the final part of the final question!), but by now you should realise that it requires the answers to the previous parts. It doesn't look related but it is (nothing is random), and demonstrates the importance of practising with textbook and past HSC questions to gain deep understanding and experience with different styles of questions.

Recognising the need to use Viete's formulas (sums and products of roots) from the Year 11 (!) Polynomials topic in a trigonometric question is the key here. Mathematics Extension 1 students should be able to identify what is needed.

HSC exam reference sheet

Mathematics Advanced, Extension 1 and Extension 2

© NSW Education Standards Authority

Note: Unlike the actual HSC exam reference sheet, this sheet indicates which formulas are Mathematics Extension 1 and 2.

Measurement

Length

$$l = \frac{\theta}{360} \times 2\pi r$$

Area

$$A = \frac{\theta}{360} \times \pi r^2$$

$$A = \frac{h}{2}(a + b)$$

Surface area

$$A = 2\pi r^2 + 2\pi rh$$

$$A = 4\pi r^2$$

Volume

$$V = \frac{1}{3}Ah$$

$$V = \frac{4}{3}\pi r^3$$

Functions

$$x = \frac{-b \pm \sqrt{b^2 - 4ac}}{2a}$$

For $ax^3 + bx^2 + cx + d = 0$:*　　　　*EXT1

$$\alpha + \beta + \gamma = -\frac{b}{a}$$

$$\alpha\beta + \alpha\gamma + \beta\gamma = \frac{c}{a}$$

$$\text{and } \alpha\beta\gamma = -\frac{d}{a}$$

Relations

$$(x - h)^2 + (y - k)^2 = r^2$$

Financial Mathematics

$$A = P(1 + r)^n$$

Sequences and series

$$T_n = a + (n - 1)d$$

$$S_n = \frac{n}{2}\big[2a + (n - 1)d\big] = \frac{n}{2}(a + l)$$

$$T_n = ar^{n-1}$$

$$S_n = \frac{a(1 - r^n)}{1 - r} = \frac{a(r^n - 1)}{r - 1}, r \neq 1$$

$$S = \frac{a}{1 - r}, |r| < 1$$

Logarithmic and Exponential Functions

$$\log_a a^x = x = a^{\log_a x}$$

$$\log_a x = \frac{\log_b x}{\log_b a}$$

$$a^x = e^{x \ln a}$$

Trigonometric Functions

$$\sin A = \frac{\text{opp}}{\text{hyp}}, \quad \cos A = \frac{\text{adj}}{\text{hyp}}, \quad \tan A = \frac{\text{opp}}{\text{adj}}$$

$$A = \frac{1}{2}ab\sin C$$

$$\frac{a}{\sin A} = \frac{b}{\sin B} = \frac{c}{\sin C}$$

$$c^2 = a^2 + b^2 - 2ab\cos C$$

$$\cos C = \frac{a^2 + b^2 - c^2}{2ab}$$

$$l = r\theta$$

$$A = \frac{1}{2}r^2\theta$$

Trigonometric identities

$$\sec A = \frac{1}{\cos A}, \quad \cos A \ne 0$$

$$\csc A = \frac{1}{\sin A}, \quad \sin A \ne 0$$

$$\cot A = \frac{\cos A}{\sin A}, \quad \sin A \ne 0$$

$$\cos^2 x + \sin^2 x = 1$$

Compound angles*

$$\sin(A + B) = \sin A\cos B + \cos A\sin B$$

$$\cos(A + B) = \cos A\cos B - \sin A\sin B$$

$$\tan(A + B) = \frac{\tan A + \tan B}{1 - \tan A\tan B}$$

If $t = \tan\dfrac{A}{2}$, then $\sin A = \dfrac{2t}{1 + t^2}$

$$\cos A = \frac{1 - t^2}{1 + t^2}$$

$$\tan A = \frac{2t}{1 - t^2}$$

$$\cos A\cos B = \frac{1}{2}\big[\cos(A - B) + \cos(A + B)\big]$$

$$\sin A\sin B = \frac{1}{2}\big[\cos(A - B) - \cos(A + B)\big]$$

$$\sin A\cos B = \frac{1}{2}\big[\sin(A + B) + \sin(A - B)\big]$$

$$\cos A\sin B = \frac{1}{2}\big[\sin(A + B) - \sin(A - B)\big]$$

$$\sin^2 nx = \frac{1}{2}(1 - \cos 2nx)$$

$$\cos^2 nx = \frac{1}{2}(1 + \cos 2nx)$$

Statistical Analysis

$$z = \frac{x - \mu}{\sigma}$$

An outlier is a score
less than $Q_1 - 1.5 \times \text{IQR}$
or
more than $Q_3 + 1.5 \times \text{IQR}$

Normal distribution

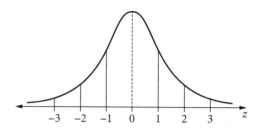

- approximately 68% of scores have z-scores between −1 and 1

- approximately 95% of scores have z-scores between −2 and 2

- approximately 99.7% of scores have z-scores between −3 and 3

Discrete random variables

$$E(X) = \mu$$

$$\text{Var}(X) = E\big[(X - \mu)^2\big] = E(X^2) - \mu^2$$

Probability

$$P(A \cap B) = P(A)P(B)$$

$$P(A \cup B) = P(A) + P(B) - P(A \cap B)$$

$$P(A|B) = \frac{P(A \cap B)}{P(B)}, \quad P(B) \ne 0$$

Continuous random variables

$$P(X \le r) = \int_a^r f(x)\,dx$$

$$P(a < X < b) = \int_a^b f(x)\,dx$$

Binomial distribution*

$$P(X = r) = {}^nC_r\, p^r (1 - p)^{n-r}$$

$$X \sim \text{Bin}(n, p)$$
$$\Rightarrow P(X = x)$$
$$= \binom{n}{x} p^x (1 - p)^{n-x}, \quad x = 0, 1, \ldots, n$$

$$E(X) = np$$

$$\text{Var}(X) = np(1 - p)$$

Differential Calculus

Function	Derivative
$y = f(x)^n$	$\dfrac{dy}{dx} = nf'(x)[f(x)]^{n-1}$
$y = uv$	$\dfrac{dy}{dx} = u\dfrac{dv}{dx} + v\dfrac{du}{dx}$
$y = g(u)$ where $u = f(x)$	$\dfrac{dy}{dx} = \dfrac{dy}{du} \times \dfrac{du}{dx}$
$y = \dfrac{u}{v}$	$\dfrac{dy}{dx} = \dfrac{v\dfrac{du}{dx} - u\dfrac{dv}{dx}}{v^2}$
$y = \sin f(x)$	$\dfrac{dy}{dx} = f'(x)\cos f(x)$
$y = \cos f(x)$	$\dfrac{dy}{dx} = -f'(x)\sin f(x)$
$y = \tan f(x)$	$\dfrac{dy}{dx} = f'(x)\sec^2 f(x)$
$y = e^{f(x)}$	$\dfrac{dy}{dx} = f'(x)e^{f(x)}$
$y = \ln f(x)$	$\dfrac{dy}{dx} = \dfrac{f'(x)}{f(x)}$
$y = a^{f(x)}$	$\dfrac{dy}{dx} = (\ln a)f'(x)a^{f(x)}$
$y = \log_a f(x)$	$\dfrac{dy}{dx} = \dfrac{f'(x)}{(\ln a)f(x)}$
$y = \sin^{-1} f(x)$	$\dfrac{dy}{dx} = \dfrac{f'(x)}{\sqrt{1 - [f(x)]^2}}\,*$
$y = \cos^{-1} f(x)$	$\dfrac{dy}{dx} = -\dfrac{f'(x)}{\sqrt{1 - [f(x)]^2}}\,*$
$y = \tan^{-1} f(x)$	$\dfrac{dy}{dx} = \dfrac{f'(x)}{1 + [f(x)]^2}\,*$

Integral Calculus

$$\int f'(x)[f(x)]^n\,dx = \frac{1}{n+1}[f(x)]^{n+1} + c$$
$$\text{where } n \neq -1$$

$$\int f'(x)\sin f(x)\,dx = -\cos f(x) + c$$

$$\int f'(x)\cos f(x)\,dx = \sin f(x) + c$$

$$\int f'(x)\sec^2 f(x)\,dx = \tan f(x) + c$$

$$\int f'(x)e^{f(x)}\,dx = e^{f(x)} + c$$

$$\int \frac{f'(x)}{f(x)}\,dx = \ln|f(x)| + c$$

$$\int f'(x)a^{f(x)}\,dx = \frac{a^{f(x)}}{\ln a} + c$$

$$\int \frac{f'(x)}{\sqrt{a^2 - [f(x)]^2}}\,dx = \sin^{-1}\frac{f(x)}{a} + c\ *$$

$$\int \frac{f'(x)}{a^2 + [f(x)]^2}\,dx = \frac{1}{a}\tan^{-1}\frac{f(x)}{a} + c\ *$$

$$\int u\frac{dv}{dx}\,dx = uv - \int v\frac{du}{dx}\,dx\ **$$

$$\int_a^b f(x)\,dx$$
$$\approx \frac{b-a}{2n}\{f(a) + f(b) + 2[f(x_1) + \cdots + f(x_{n-1})]\}$$
where $a = x_0$ and $b = x_n$

*EXT1, **EXT2

Combinatorics*

$${}^nP_r = \frac{n!}{(n-r)!}$$

$$\binom{n}{r} = {}^nC_r = \frac{n!}{r!(n-r)!}$$

$$(x+a)^n = x^n + \binom{n}{1}x^{n-1}a + \cdots + \binom{n}{r}x^{n-r}a^r + \cdots + a^n$$

Vectors*

$$\left|\underset{\sim}{u}\right| = \left|x\underset{\sim}{i} + y\underset{\sim}{j}\right| = \sqrt{x^2 + y^2}$$

$$\underset{\sim}{u} \cdot \underset{\sim}{v} = \left|\underset{\sim}{u}\right|\left|\underset{\sim}{v}\right|\cos\theta = x_1 x_2 + y_1 y_2,$$
where $\underset{\sim}{u} = x_1\underset{\sim}{i} + y_1\underset{\sim}{j}$
  and $\underset{\sim}{v} = x_2\underset{\sim}{i} + y_2\underset{\sim}{j}$

$$\underset{\sim}{r} = \underset{\sim}{a} + \lambda\underset{\sim}{b}^{**}$$

Complex Numbers**

$$z = a + ib = r(\cos\theta + i\sin\theta)$$
$$= re^{i\theta}$$

$$\left[r(\cos\theta + i\sin\theta)\right]^n = r^n(\cos n\theta + i\sin n\theta)$$
$$= r^n e^{in\theta}$$

Mechanics**

$$\frac{d^2x}{dt^2} = \frac{dv}{dt} = v\frac{dv}{dx} = \frac{d}{dx}\left(\frac{1}{2}v^2\right)$$

$$x = a\cos(nt + \alpha) + c$$

$$x = a\sin(nt + \alpha) + c$$

$$\ddot{x} = -n^2(x - c)$$

*EXT1, **EXT2

9780170459259